Raising Goats for Beginners

The Complete Guide to
Breeds · Housing · Fencing
Health and Diet · Breeding
Dairy · Meat · Fiber

Max Barnes

© COPYRIGHT 2021 MAX BARNES - ALL RIGHTS RESERVED.

The content contained within this book may not be reproduced, duplicated or transmitted without direct written permission from the author or the publisher.

Under no circumstances will any blame or legal responsibility be held against the publisher, or author, for any damages, reparation, or monetary loss due to the information contained within this book. Either directly or indirectly.

Legal Notice:

This book is copyright protected. This book is only for personal use. You cannot amend, distribute, sell, use, quote or paraphrase any part, or the content within this book, without the consent of the author or publisher.

Disclaimer Notice:

Please note the information contained within this document is for educational and entertainment purposes only. All effort has been executed to present accurate, up to date, and reliable, complete information. No warranties of any kind are declared or implied. Readers acknowledge that the author is not engaging in the rendering of legal, financial, medical or professional advice. The content within this book has been derived from various sources. Please consult a licensed professional before attempting any techniques outlined in this book.

By reading this document, the reader agrees that under no circumstances is the author responsible for any losses, direct or indirect, which are incurred as a result of the use of the information contained within this document, including, but not limited to, — errors, omissions, or inaccuracies.

Your Free Gift

I'd like to offer you a gift as a way of saying thank you for purchasing this book. It's the eBook called 20 Delicious Goat Meat, Milk and Cheese Recipes. While this book contains information on how to make dairy products from goat's milk and some goat meat recipes, I made this cookbook for people who would like to do more with their goat produce and try new recipes. I hope you enjoy it! You can get your free eBook by scanning the QR code below with your phone and joining our community. Alternatively, please send me an email to maxbarnesbooks@gmail.com and I will send you the free book.

SPECIAL BONUS!
Want this book for free?

Get FREE unlimited access to it and all of my new books by joining our community!

Scan with your camera to join!

Goat Record Keeping Log Book

Keeping records of your goat herd is important and I highly recommend all new goat owners do that. Keeping records of your goats helps keep things organized and allows to take better care of your goats. I have created a Goat Record Keeping Log Book that will allow you to keep all the important information about your herd in one convenient place. If you're interested in getting a log book, please scan the QR code below with your phone to find out more. Alternatively, please send me an email to maxbarnesbooks@gmail.com and I will send you the link to the log book.

Scan with your camera to find out more

Contents

Introduction 11
    What This Book Will Cover 13

    Why I'm Writing This Book 14

Chapter 1: Getting to Know Goats 17
    The Joys and Benefits of Keeping Goats 17

    Best Goat Breeds for Milk, Meat, and Fiber 21

    Goat Terminology 28

    Registration 31

    Pedigree 33

    Facts and Myths About Goats 35

Chapter 2: Getting Your Goats 39
    Where to Buy Goats 39

    Spotting a Good Goat 42

    Getting Your Goats Home 44

    Assessing a Goat's Worth 44

    Approximate Costs of Starting a Goat Herd 45

Chapter 3: Housing Your Goats 52
    Goat Shelter Considerations 52

    Size Requirements 53

    Manure Management 54

Dry Floors 54

Bedding 55

Protection from Extreme Weather 57

The Manger 58

Additional Facilities for Breeding and Milking 61

Helpful Additions 61

Safety Considerations 62

Chapter 4: Fencing 65

Goat Fencing Considerations 66

Types of Goat Fencing 67

Gates and Latches 75

How Much Fencing Is Necessary to Protect Your Goats? 76

Chapter 5: Feeding Your Goats 78

Goats' Digestive System 79

Goats' Nutritional Requirements 80

Chapter 6: Goat Care and Grooming 90

Goat Grooming Supplies 90

Brushing 90

Hoof Care 91

Tattooing 93

Shearing 94

Leash Training 95

Chapter 7: Goat Health 98

    Basic Healthcare Requirements 98

    Vaccinations 98

    Deworming Goats 99

    Castrating Goats 101

    Disbudding and Dehorning 103

    Common Goat Health Issues 107

    How to Tell If Your Goat Is Sick 116

    Goat First Aid Kit or Medical Box 117

Chapter 8: Breeding Goats 121

    Preparing for Breeding 122

    Caring for Pregnant Goats 125

    Birthing Kids 126

    Newborn Check and Care 128

    Raising Kids 129

Chapter 9: Milking 133

    Milking Essentials 134

    Preparing to Milk 135

    Milking Procedure 135

    Pasteurizing Goat's Milk and Keeping It Fresh 136

    Cleaning Your Equipment 140

Chapter 10: Harvesting Goat Meat 142

When to Butcher a Goat 143

How to Butcher a Goat 145

Meat Quality 146

Storing Goat Meat 147

Chapter 11: Additional Benefits of Goats 149

Fiber 149

Weed Control 154

Manure 156

Goats as Pack Animals 158

Goats as Pets 159

Chapter 12: Goat Milk and Meat Recipes 162

Cheesemaking 162

Dairy products 164

Goat Meat Recipes 169

Conclusion 174

Herd Management Calendar 177

Resources 183

Index 189

Introduction

Goats are incredible creatures. They do not require much themselves, yet raising them can help you achieve a sustainable lifestyle. Goats are very versatile as animals to raise because they can produce milk, meat, fiber, help clear your land, make manure, and much more. If you want to start small, you can get a couple of goats for your backyard and develop this slowly into a smallholding or a small farm. Farming is hard work, long hours and demanding, so by starting small with a few goats, this gives you the opportunity to see if you would like to expand upon this gradually, rather than just purchasing a herd of 50 goats immediately. Goats are amazing, they can help you make money, save you money, and can be good for your health too! If people who are feeling depressed are placed in a field with goats, their playful gamboling nature can cheer them up and improve their mood.

Many people would love to keep goats, for many of the reasons mentioned above. But often people also don't know where to start. Beginners can feel overwhelmed when they start to think about keeping goats on their property because they don't have the knowledge they need, and can't necessarily trust Internet sources to be accurate and high quality. They may have heard stories of people who have tried raising goats in the past and failed, and paid out money in vet's bills, or never made a profit off their livestock. This book, however, is thoroughly researched, as well as based on sound and successful experience, and it will save you hundreds of hours of time trying to do the research yourself, having all the information you need in one handy place. It will give beginners everything they need to know about keeping goats for milk, meat, or fiber and how to make a successful living out of doing so.

What This Book Will Cover

This book has twelve chapters that will give you a complete guide for beginners, with everything you need to know about raising goats.

Chapter 1 will introduce you to goats, goat terminology, different goat breeds and which are best for specific purposes, such as milk, meat, or fiber, as well as myths and truths about goats. This will help you become familiar with terminology so that you understand goat terms when reading about the topic or speaking with others. Every subject has terminology that is specific to it, whether that's biology, engineering, English literature, and so on, and the same can be said with goats.

Chapter 2 will look at where you can buy goats from, and how to make a good purchase, then transporting your goats to your home/farm. This chapter will help to ensure that the goats you buy suit your needs and are the healthiest goats you can buy. It will also enable you to get them home safely.

Chapter 3 will look at the shelter you require for goats, and its various considerations, such as size, bedding, protection from the elements, and much more. This will help you provide adequate accommodation and bedding so that your goats are safe and well cared for.

Chapter 4 will look at the fencing you need to protect your goats. This will give you the best tips and tricks about how to select the correct fencing, the amount you'll need, and keep your goats where they should be on the property.

Chapter 5 focuses on what to feed your goats to ensure they get a nutritional diet and stay healthy, and how to formulate a goat ration. Like with humans, how healthy a goat is depends a lot on what it is fed. You want to ensure each goat gets a fair share of food and that they're not overfed.

Chapter 6 is about caring for goats and grooming them, care of hooves, hair, dehorning, disbudding, and tattooing them. This chapter will help you learn how to best take care of your goats, to keep them in good health, happy, and safe.

Chapter 7 looks at goat health, veterinarians, and common health issues to be aware of. There are certain common diseases and illnesses that can affect goats, like with any animals. But, by being well informed about these, you can spot them early and get them treated by a vet if they occur.

Chapter 8 is about breeding goats, preparing goats for breeding, caring for pregnant goats, birthing goats, and raising kids. If you want to produce goats to sell to others, or for meat, this chapter will be important to you and give you the information you require.

Chapter 9 is about milking goats and the equipment you will require. You can learn many tips on how to produce the best tasting goat's milk in this chapter.

Chapter 10 is about harvesting goat meat, butchering, and storing meat. Learning the right places to store fresh and frozen meat is essential and considering purchase times and destinations can be important too.

Chapter 11 looks at the additional benefits of having goats, such as fiber, weed control, manure, using goats as pack animals, and as pets. This may really help you think outside the box with goats, and not just consider meat and milk production as there truly are so many other benefits of having goats.

Chapter 12 looks at recipes that use goat's milk or goat meat. Goat's milk and cheese are delicious and can really add a delicious flavor and texture to your meals. Goat's milk is popular too, and this chapter will give you some recipe inspirations to try.

Why I'm Writing This Book

My name is Max, and I grew up on a farm helping my grandmother Anna raise her animals. One of my earliest memories was picking up eggs from chickens into a basket. My grandmother had a wide range of farm animals, but from being young I really loved goats, they were my favorite animals, because they were so friendly and came to me to cuddle. I loved everything about them, from their eyes, their mischievousness, their energy, and quirky personalities.

I really enjoyed helping my grandma feed the goats, take care of them and milk them. I remember being given glasses of delicious goat's milk to drink in the afternoon, with one of my grandma's homemade cookies. Goat's milk and cheese is a super-popular alternative to cow's milk, which many people find they have allergies to.

Nowadays, I have a farm of my own, and keep various animals on my property, the majority being goats. This helps me live a sustainable lifestyle, producing milk, meat, fiber, and manure, but also every day I'm reminded of fond memories from my childhood, where I helped my grandmother with her farm. I learned everything I know from my grandma about raising goats and other animals. But I've also added to this learning over time by researching, networking with other farmers, the staff I've worked with, attending farmer's markets, and by the practical day-to-day experiences.

I think that everyone has something they are good at and can teach others; and my knowledge is definitely focused on goats. One of the key reasons why I've written this book is because I want to show people that raising goats is not as difficult as some people may think, and the benefits far outweigh the time, money, and effort invested.

So, this book will teach you what you need to know about raising goats to produce milk, meat, fiber, help clear your land, and make manure. You will learn about breeding and raising goats. It is not a guide to raising any other animals apart from goats.

So, if you are ready to take the next step in becoming knowledgeable and informed about goats, read on to Chapter 1, which is all about getting to know goats.

Chapter 1: Getting to Know Goats

If you are interested in getting goats, it is still important to think really carefully about this, do your research, read the chapters in this book to give you true insight, and be a responsible goat owner. You are essentially taking on the responsibility for multiple lives, and you need to ensure you treat them as well as you possibly can. Like people say about dogs, "A dog is for life, and not just for Christmas", goats are for their lives, and not something done impulsively. The fact that you've purchased this book is a good sign, showing that you're interested in doing it properly, and you're in it for the long haul. Goats typically live for 8–14 years (similar to a dog or a cat). You need to think about who will look after your goats when you want to go on holiday, or if there is an emergency or something unexpected crops up. It can be trickier to find someone to look after goats, rather than a dog or cat sitter. So as not to spread any diseases amongst farms, it is important to not wear your typical farming clothes when you visit other people's farms and request that they do the same.

You need to have thoroughly thought about all aspects covered in this book before getting goats, such as where you will keep them, where you will buy their feed from, and which veterinarian will look after them for you if they need health care.

Once you start to purchase and look after goats, it is really hard to stop. There are so many different varieties that can add so many benefits to every aspect of your life. Goats are fun, vivacious, mischievous creatures and will fill your life with joy and entertainment, as well as having the ability to create a profitable, sustainable life.

The Joys and Benefits of Keeping Goats

There are numerous benefits to keeping goats. They are fairly inexpensive to purchase, look after, house, and feed. They will eat food scraps, which means you can minimize wastage. They're easy to handle, being placid in nature. The actual goat skin can be dried and tanned in a similar manner to leather and produce things like gloves. In Africa, their hide is used for drums. Rugs can be made from goatskins too. Adults can enjoy the social benefits that owning goats can involve, if you decide to get involved in exhibiting them at national shows. Goats can give children wonderful learning opportunities away from computers, phones, and other IT distractions, connecting them with nature, giving them the opportunity to care and develop empathy, getting them working as part of a team to care for them, and giving them responsibility and commitment. I will forever be grateful for the opportunities I had learning about goats from my Nana, it was special time that I spent with her that I enjoyed so much and look back on fondly. A few sources suggest that you can combine goats with yoga classes! So, if you are a yoga instructor and would like to have goats roaming around as people do their yoga, this will surely make the classes more entertaining and bring people closer to nature. You could use goats for small treks to carry light loads, and they won't need food packing for themselves, as they can graze on the way.

Some other joys and benefits are explored further below:

1. Goats can do the gardening for you

Goats will eat the weeds on your property, they will eat most things, including poison ivy, blackberry bushes, kochia, yellow star thistle, wild turnip, wild rose, spotted knapweed, and any overgrown plants. This saves you the job of doing it and saves you from needing to use weedkillers, which can be damaging to the environment. Goats can also eat leaves, bark, and pine needles. Goats enjoy having food to browse, and they don't choose to eat grass unless there is nothing else for them to eat, so they are unlikely to keep your lawn looking nice. Goats can be used to clear land, which reduces fire risk, and can reduce weeds on public land and around schools. They can also be good near streams to prevent brambles from choking the stream. You can charge people to let your goats clear their weeds. By getting goats to clear your land for you, you are saving money on hiring land clearance workers to do this for you or hiring/renting expensive equipment to do the job.

2. **Rich manure**

Goats produce a manure rich in nitrogen and phosphate, which is perfect for keeping garden plants, vegetables, and flowers blooming beautifully. They also contain potassium and potash and other minerals. Goat pellets are dry, which makes them easy and non-messy to collect, and once they have completely dried out, they don't smell, unlike cow manure. You can also burn goat manure as a fuel on fires, so if you want to live a self-sufficient lifestyle this could be an important use.

3. **Make soap**

Goat's milk can be used to make a very mild soap, that is perfect for people who have sensitive skin. The soap is very soft and gentle.

4. **Goats as pets**

If you are purchasing goats solely because you want them as pets, buying two males is a good combination. Goats are social creatures and like the company of other goats, it would not be fair to just purchase one. Pet goats can include neutered males (known as wethers), these are less expensive than does for milking, or bucks (unneutered male goats) which are required for breeding.

Goats make wonderful pets, they are incredibly gentle around children, they show loyalty and have a really fun, playful side that is entertaining to watch. They have a great ability to jump and spring about injecting energy and vitality into the environment. Goats can be used for pet therapy to give comfort to others, and they do enjoy being petted. If you are buying goats as pets for children, pygmy goats can be a good choice, because they are small and perfect for children. If you want your goats more for pets, rather than for milk or meat, you could buy goats and then open a petting zoo so that people can pay a small fee to come and stroke the goats.

5. **Delicious milk and cheese**

Dairy goats can mean that you have your own supply of milk and cheese at home, and no longer have to purchase these from a supermarket. Having a goat, rather than a cow, is cheaper, they're smaller and take up less space. If you care lactose intolerant, goats' milk is easier to tolerate and digest than cow's milk. You can make incredible cheese from goat's milk, which is called chevre, it has a tangy distinct taste and often people put honey with it. The types of goats that are well suited for producing cheese and milk include Alpine, Saanen, and Nubian goats.

Alpine goat    *Saanen goat*    *Nubian goats*

Best Goat Breeds for Milk, Meat, and Fiber

Best Breeds for Milk Production

If you decide to get goats to milk, this is very much a lifestyle. It is time intensive, and you probably won't plan many vacations if you have goats to milk. Dairy goats provide a plentiful supply of milk, and it is likely to become more than you can use as a family. An average dairy goat will produce 6–8 pounds per day. But a high yielding doe may produce 16 pounds of milk per day.

Does need to be bred in order to produce milk. Once your goat is pregnant, she will produce milk right after giving birth to her first kid. The lactation period can last from months to years depending on the breed. The average lactation period for dairy goats is 284 days (about 9 and a half months), and peak production usually occurs 4–6 weeks after kidding (giving birth).

If you have more milk than you can use, this is when you can also make a wide variety of other products from the goat's milk: cheese, yoghurt, kefir, fudge, gelato, to name but a few. If you'd like to keep goats for milking, but don't have a lot of space, Nigerian dwarf dairy goats can be used to produce incredible cheese, yoghurt, kefir, fudge, and gelato. Nigerian dairy goats are a desert breed, which means their milk has a high fat content, and it contains a lot of protein too. Another advantage of this type of goat is that because the goats are so small, they only take up one third of the space that full-sized goats would usually need. So, this is definitely something to consider if you don't have the space for full-sized goats.

Goat's milk is tolerated better by those who are lactose intolerant because it's easier to digest than cow's milk. Goat's milk can be of high value to elderly people, sick people, babies, and those with an allergy or intolerance to cow's milk. If you are trying to provide a milk substitute to an orphaned animal, such as a foal or puppy, goat's milk is the best substitute for this too. Goat's milk is higher in vitamin A, choline, and inositol than cow's milk, but is lower in vitamins B6, B12, Vitamin C and carotenoids. There are less goats than cows in the US though, which means sometimes goat's milk can be harder to find, and therefore produces a more profitable income with higher prices than cow's milk. If you want to sell food products, make sure to do research into local laws regarding this, as each State in the US does vary.

The American Goat Federation said that in 2017 the US had 373,000 goats for milk, with most found in Wisconsin, then California. Below are some suggested good breeds for dairy goats, later Chapter 9 will cover everything you need to know about milking goats: what equipment you'll need, the milking procedure, handling milk and preparing it for human consumption, how to keep it fresh, plus things like sterilizing equipment to keep it safe and hygienic. The key goats that the American Goat Federation suggests for dairy goats include:

- **Alpine Goats:** these goats are known for good health and excellent high-volume milk production. They originated in French Alps and they can thrive in almost any climate. Alpine Goats are often considered to be the highest producing milkers, top goats can produce up to 2 gallons per day.

- **LaMancha Goats:** these goats have a nice temperament and produce milk that is high in butterfat. The main advantage is that their lactation period can last up to 2 years after kidding. These goats are easy-going and really sturdy.

- **Nigerian Dwarf Goats:** these originate from West Africa, but are now popular in the US. They may be small, but they usually produce about a quart of milk per day. Nigerian Dwarf Goats are a great choice if you don't have a lot of space.

- **Saanen Goats:** these goats originated in Switzerland. They are typically white or light cream in color, although white is preferred. They are the largest dairy goats, and are second in milk production only to Alpine goats. These are a popular choice for dairy goats and they generally have an easy-going temperament.

[1] Image from Britannica: https://www.britannica.com/animal/LaMancha

- **Sable Goats:** these goats are derived from the Saanen breed. They are medium to large in size and, just like Saanen goats, they are great milk producers. They look very similar to Saanen goats, expect for coloration. They can be any color or combination of colors, except for white and light cream.[2]

- **Toggenburg Goats**: these goats originated in Switzerland in Toggenburg Valley. It was the first breed of goat to receive star status. They are medium in size and moderate in terms of milk production. These sturdy goats are known as being the oldest dairy goat breed.

- **Nubian Goats**: these goats have a peculiar appearance, and their milk is high quality and high in butterfat. They are easily recognizable due to their long, floppy ears. They originated in the Middle East, so they thrive in hotter climates, and they have a longer breeding season. Although they may not produce as much

---

[2] Image from allaboutgoats.com: https://allaboutgoats.com/sable-goat/

milk as other breeds, it is very high in fat content. Nubian goats are great for meat too, and they are typically raised for both meat and dairy.

- **Oberhasli Goats:** these goats originated in the Canton of Berne in Switzerland. The typical color of goat is a reddish brown. It produces lovely, sweet tasting milk. These are medium-sized goats which are great as pack animals due to their strength and sure-footedness. They are still relatively rare in the United States, so you might have a hard time trying to find these goats for purchase.[3]

Best Breeds for Meat

Goat meat is also called chevon. It is a red meat that is gaining in popularity, especially in the US where it is frequently eaten by Caribbean, Hispanic, Muslim, and Chinese customers. Goat meat is a similar texture to venison. It contains less fat than chicken, yet more protein than beef and is therefore a healthy meat. It contains more iron than beef, lamb, chicken, and pork. Goat meat provides a good source of conjugated linoleic acid, which is good to aid fat burning, weight loss, metabolic health, and a lower risk of other diseases. It is a tasty meat, and very tender.

Raising your own supply of chevon, or cabrito (from young kids) ensures that you have it ready at hand. In Chapter 12 of this book, you will find delicious recipes that use goat's milk, cheese, and meat. When you raise your own goats, this can mean you have plenty of meat for family meals. You can also have a small farm business providing fresh or frozen goat meat to others, which can generate income for you. When you have raised your own meat, you know that the animal was well treated, that the animal didn't have any diseases, and was kept in clean

---

[3] Image from WNC Nature Center: https://wildwnc.org/animals/oberhasli-goat/

conditions. Goat meat prices are on the rise at the moment, in 2021 a pound of goat meat typically cost somewhere between $3.50 and $3.80, up from $2.50–$2.75 5-year average.

These goat breeds are especially good for meat:

- **Boer Goats**: these arrived in the US in 1993, they grow quickly, they are docile in nature and have high fertility. They are characterized by the red head and red on at least a portion of the neck. They have large floppy ears. Boer goats are the largest goat breed and they tend to be in high demand because they grow fast and produce great carcasses.

- **Savanna(h) Goats**: this type of goat reproduces well, has good muscles, bones, and strong legs and hooves. Their kids also develop faster than other goats. It is common for this type of goat to give birth to twins.

- **Kiko Goats**: these originated in New Zealand and the word "kiko" means meat in Maori. They are known for their hardiness and their ability to gain substantial weight under natural conditions without supplements. Kiko goats are aggressive foragers and can thrive in less than ideal conditions.
[4]

- **Myotonic Goats**: these are often called fainting goats. They are placid and friendly. They don't tend to climb or jump over fences, which can make looking after their enclosure easier. The kids are easy to birth, the does are fantastic protective mothers, and they can breed all year round. They are called Myotonic, based on a medical term that describes stiffness called Myotonia congenita, if the goats become startled, or get too excited, their muscles stiffen and they tip over on their sides. This isn't an issue, and the goat is not harmed, the goat remains conscious when this happens.

---

[4] Image from livestockpedia.com: https://livestockpedia.com/goats/kiko/

Best Breeds for Fiber

Some goats produce wool. If you want to produce cashmere wool, then you need Cashmere goats. If you want to produce mohair, then you need Angora and Pygora goats to produce mohair. Angora goats originally came from Asia Minor. With the wool that you gain from the goats, there is a wealth of products you could choose to make, from jumpers to blankets, scarves, hats, and so on. You could knit, crochet or weave with the goat's wool. If you are not into knitting yourself, you can still benefit by selling the wool to local companies. Cashmere goats are known for their wool being incredibly soft and luxurious, and therefore cashmere and Angora products are usually quite expensive.

Cashmere goat    *Angora goat*

Goat Terminology

Listed below are some common terms that are often used when discussing goats. If you haven't heard of them before, don't worry, you'll soon become familiar with them. People who have been around goats a long time will use these terms freely, without necessarily explaining because they are used to them. The glossary below may help.

Goat Types

Billy: This is the name for a male goat that is usually uncastrated. Sometimes this name can be frowned upon in the goat community, and people prefer the word 'buck' (see below). Billy is considered an old-fashioned outdated term now.

Bucklings: A buckling is a young male goat that has not been castrated but is not old enough yet to be a buck and to breed.

Buck: This is the name for an unneutered male goat, perfect for breeding. Most professional people will call male goats this (they won't use the term 'billy').

Doeling: This is the name for a young female doe, that is not yet ready to breed.

Does: These are female goats good for producing milk, which in turn can be used to create yoghurt, cheese, butter, gelato, fudge, kefir, and other dairy products.

Heard Queen: The dominant female goat in the herd. This goat leads the herd. She remains in charge until she is too old or ill, then one of her daughters usually replaces her at that point.

Kids: This is the name for baby goats or young goats.

King: This is the name for the dominant male goat (a buck) in the herd at breeding time.

Nanny: This is a name for a female goat, but the word 'doe' is much preferred. The word 'nanny' is considered old fashioned and outdated now.

Wethers: This is the name for a neutered male goat. They tend to be less expensive than does or bucks, and make great pet goats. They can't breed and having them castrated will prevent buck behavior and mating. As well as for pets, this type of goat can be used for meat.

Goat Health

Abscess: This is a boil filled with pus. It can be an indication of CL.

Banding: This is a way of fixing a male goat.

CDT: This is a vaccine used annually to protect goats against Clostridium perfringens type C + D as well as tetanus.

CL/CAE/TB/Johne's/Brucellosis: All diseases that goats can get, most of which can be tested for to make sure your herd is clean.

Cocci: A sickness that can cause diarrhea.

Drenching: This is when you dose a goat with a liquid medicine.

FAMACHA Scoring: This is a scoring based on the color of a goat's eyelid, which indicates how anemic they may be.

Fixed: This is another word for when an animal has been castrated/neutered.

Intact: This is another way of saying that a male goat has not been neutered or castrated.

Scours: This is basically diarrhea.

Goat Anatomy

Beard: Male goats or bucks have beards, hair growing under their chin.

Cleat: A goat's hoof is different from a horse's hoof and is split into two parts like a deer. Each half of a goat's hoof is known as a cleat.

Cud: The regurgitated food that goats chew, swallow, and digest at this point.

Disbudded: This is a term used when a goat has had its horns removed when it was very young.

Horns: Goats' horns are very different from deer antlers. Horns have blood vessels in them, and are not shed like antlers.

Poll and Polled: The poll is the space on top of the goat's head between the horns. A polled goat is one that is born without horns, and some goats are deliberately bred this way to avoid horns.

Ruminant: This is the term for animals that have four compartments in their stomach, like goats and cows. The four compartments are called rumen, reticulum, omasum, and abomasum. These animals regurgitate their food and chew it as cud (see above definition) to help digest it.

Scurs: These are small pieces of horn that have grown back, or horns that were not fully removed when a goat went through disbudding (see above for term).

Udder: This is the part of female goats that hangs between their back legs; it is their mammary gland.

Wattle: Some goats have wattles, which are bits that hang from their neck, covered in fur. There is no purpose to these, they appear to be decorative.

Goat Reproduction

Freshen: This means that a goat is starting to produce milk. A first freshener is a goat who is producing milk for the first time.

Kidding: When a goat has babies.

Open: This term is used when a female goat is not pregnant.

Rut: This is when bucks are ready to breed. They will go through periods of time when they have a lot of hormones raging.

Weaned: This is when a baby goat no longer needs to nurse from its mother.

Goat Produce

Cabrito: This is another name for goat meat, it usually refers to a young goat's meat (like veal in comparison to venison).

Chevon: This is the name for goat meat.

Chevre: This is the name for cheese made from goat's milk.

Registration

There are quite a few of different registries and associations in the US. Many people think that you only need to register your goats if you want to show them, but that's not entirely true. There are a lot of benefits to registering goats. Registries have listed breed standard requirements, and animals that don't conform to the standards can't be registered. This helps prevent unwanted characteristics or problems from developing. Typically, rules for goat registration are fairly universal. In most cases, only goats with registered purebred parents can be registered. Also, if an animal is purchased, a transfer of ownership must be submitted before the new owner can register the goat. If the goat was already registered before purchase, then only the transfer of ownership is required.

Purebred goats are more expensive, of course, but there are multiple benefits to buying purebred animals. Unregistered goats can be poor quality. Reputable breeders won't sell a goat with papers if it has a defect. Registered goats often have records, evaluations and tests. So, when buying goats from a breeder, you can be sure you're buying a healthy purebred goat. While the initial cost may be higher, keep in mind that the cost of feeding and taking care of registered and unregistered goats is the same. However, kids of registered goats can be sold for more.

With that said, American Dairy Goat Association (ADGA) is the largest dairy goat registry in the United States. It includes Alpine, LaMancha, Nigerian Dwarf, Nubian, Oberhasli, Saanen, Sable, and Toggenburg breeds. Membership is not required for goat registration, but it lowers the registration fee and provides many other benefits.

The American Goat Society (AGS) is a smaller organization that accepts the following breeds: French Alpine, LaMancha, Nigerian Dwarf, Nubian, Oberhasli, Pygmy, Saanen, Sable, and Toggenburg.

If you're in Canada, you can register your goats with The Canadian Goat Society. They maintain the herd books for nine purebred breeds of goats and a Special Registry for upgraded animals as specified under the animal pedigree act, which are: Alpine, Angora, La Mancha, Nigerian Dwarf, Nubian, Oberhasli, Pygmy, Saanen, and Toggenburg.

There are quite a few other goat associations and societies. Now, with the ubiquitous internet access, I would recommend you do your own research and find an organization that suits your needs.

Pedigree

Here is the chart with milk production and pedigree abbreviations for ADGA and AGS:

Designation	ADGA	AGS	Requirement
Champion	CH	MCH	3 Grand Champion show wins, judged by at least 2 different judges.
Star Doe	*M	*D	A doe that has met the minimum requirements for milk production or has 3 *D/M daughters or 2 +S/B sons, or 2 *D/M daughters and 1 +S/B son.
Two Star Doe	2*M	2*D	A doe that is the daughter of a Star Doe and has also met the minimum standards for milk production. The number before the star indicates the number of consecutive generations of qualifying does.
Star Sire	*B	*S	A buck that has a *D/M dam and +S/B sire or sire with a *D/M dam.
Plus Sire	+B	+S	A buck that has at least 3 *D/M daughters from different does or has 2 +S/B sons, or has 2 *D/M daughters and 1 +S/B son.
Two Plus Sire	++B	++S	A buck that has at least 3 *D/M daughters from different does and at least 2 +S/B sons.
Two Plus Star Sire	++*B	++*S	A buck that has at least 3 *D/M daughters, 2 +S/B sons and a *D/M dam or +S/B sire.
Permanent Grand Champion	GCH	ARMCH	A goat that has achieved Champion status and also earned a milk production star.
Superior Genetics	SG	—	Identifies a doe or buck that is in the top 15% of the production index for that breed.
Superior Genetics Permanent Grand Champion	SGCH	—	Identifies a doe or buck that has earned both Superior Genetics and Permanent Grand Champion status

Pedigrees can be confusing at first sight, and to people new to goats they look like a bunch of random letters and signs. Designations may vary between registries; however, letters before the goat's name typically signify its championship status, and sometimes milk status. The letters after the name signify milking records and classification or linear appraisal scores. Names consist of two parts. The first part of the name will be the herd name—the farm that bred the goat. This will be followed by the goat's name.

In addition to the milk production and show awards in the chart above, ADGA and AGS use classification systems to judge the conformational quality of a goat.

AGS uses a classification system to rate goat conformation which compares the goat to an ideal 100% and assigns a percentage for that goat. The scores are Excellent (90–100), Very Good (85–89.9), Good Plus (80–84.9), Good (70–79.9), Fair (60–69.9), and Poor (under 60). This classification is displayed after the goat's name and any production awards.

ADGA uses linear appraisal to classify goats with the goat being assigned scores for general appearance, dairy character, body capacity, and mammary. The classifications are Excellent (E), Very Good (V), Good Plus (+), Acceptable (A), and Poor (P). These designations are generally placed below the goat's name.

The linear appraisal scores or classification score for a goat tend to increase as they age until they are somewhere between four to seven years old, so lower scores for younger goats are typical. That's because it's difficult for young does or bucks to measure up conformationally to mature older does and bucks.

Let's take a look at an example of a doe pedigree designation:

SG GCH Mountain Lake Emma 3*M VEEE90

Let's go through it step-by-step. SG stands for Superior Genetics and indicates that she is in the top 15% of the breed for the PTI. GCH signifies that she is a grand champion. The first part of the name (Mountain Lake) is the herd name—the farm that bred the goat. This is followed by the goat's name—Emma. Next is the milk production award. The 3 before the * indicates that she is a third-generation star earner. And last but not least is the ADGA Linear Appraisal Score of VEEE90. V's and E's represent different structural categories she is rated on, such as general appearance, dairy character, body capacity, and mammary. V stands for Very Good and E is for Excellent.

Now that you know what all the letters, numbers and symbol mean, you should be able to better understand a goat's pedigree. Also, if you look at the whole pedigree including the goats dam, sire, grand dam, grand sire, and so on, you will be able to see what traits trend in their ancestry. This will hopefully allow you to choose goats that suit your needs and goals with your herd.

Facts and Myths About Goats

Facts

- Goats are ruminants and have four stomach compartments. They chew cud. Their four stomachs enable them to digest, regurgitate, and re-digest their food. Goats are herbivores, meaning they eat plants, herbs, and tree leaves. They swallow their food without much chewing, then they regurgitate it (it is called a cud at this point), then chew it thoroughly before swallowing it to be digested.
- Goats only use one side of their mouths to grind food, which causes a rotary movement when chewing.
- There are two types of goats, domestic goats (Capra hircus) and mountain goats (Oreamnos americanus). It is the domestic goats that are used as farm animals.
- Goats have 60 chromosomes (humans have 46).
- You don't have to shear or comb goats, like you do with sheep. But Angora goats can be sheared to produce mohair fiber.

- Goat horns are narrow and straight; sheep horns curl in loops at the side of their heads.

- In bright light, the pupil of a goat's eye is rectangular, rather than round.

- Goats live in herds, usually of around 20 goats in the wild. They are more social during winter, whereas in the summer they will wander off solo. There is a dominant female in the group (known as the herd queen) until the breeding season, when a male dominates.

- Did you know that two Presidents of the United States have had pet goats at the White House? This included Benjamin Harrison and Abraham Lincoln!

Myths

Below are some things people believe about goats, but these are actually incorrect:

- **Goats will headbutt you**: Goats will butt one another because this is how they develop dominant goats within a herd. But, if goats are trained as kids, from an early age, they won't butt humans when they grow into adult goats. If a kid butts against your legs when it is young, the advice is to never push back! Goats will only be aggressive towards humans if you've raised them to be. Goats need lots of space when they are young.

- **Goats eat anything**: This is not true, and goats browse foods and can be selective. They use their mouth to touch and feel, so sometimes they are just investigating things, but not actually eating them. Goats will eat rose bushes and lilac bushes, but they won't eat tin cans or rubbish.

- **Goat's milk tastes off**: Goat's milk can absorb smells of the surrounding area or be influenced by the food that goat has eaten. If the goat has eaten lots of wild garlic, the milk may taste of this. Mint and onions can affect the flavor of goat's milk too. But, provided you keep the area you are milking the goats in is super clean, and if you carefully monitor their diet, your milk should be delicious.

- **Goats smell**: Male goats in breeding season will have a musky smell, and this will rub off on female goats. But this happens only during the breeding season. Goats do not smell at other times of the year.

- **Only male goats have beards and horns**: Male and female goats can have both beards and horns.

- **Goats maintain lawns**: This is not true. Goats prefer to browse, rather than graze. If you placed them in your garden, they will eat from bushes, trees, flowers, plants, and probably ignore the grass.

- **It is safe to tether goats**: I wouldn't ever advise anyone to do this and leave the goat unwatched. It can be dangerous, the goat can become tangled, or even hang itself.

We now have lots of different breed of goats, and utilize them for everything: meat, milk, fiber, we use them as land-clearers, and to make a wide variety of products including various dairy products, soap, and even clothes. Our whole life revolves around goats and having them is what enables us to live freely, according to our own rules. We work hard, but it is immensely rewarding to be so self-sustainable. We eat goat meat, we drink goat's milk, we eat goat cheese, eat ice cream and yoghurt made from our goat's milk. We make clothing and blankets from goat fiber. We fertilize our plants, vegetables, and other parts of the garden with goat manure. We use goat manure to start a fire if we need to, and burn it in winter. Goats have now become central to our way of life, and they could be to yours too. You don't have to start big if you don't wish to. You could start small, with a couple of goats, and then grow from there if this way of life suits you.

By reading this chapter, I hope it has encouraged you to see the many benefits that keeping goats can bring. I hope the glossary of terms has made you feel more confident when reading and speaking about goats and now you can use and understand the correct terminology connected to them. Finally, I hope you found the facts interesting, and have had any myths you may have heard about goats dispelled.

This book won't promote any specific type of breed, that is entirely your choice, but it is important to choose goats that are healthy and free from disease. The next chapter, Chapter 2, will look at getting your goats, where to buy them, the cost, choosing good goats, and getting them home in more detail.

Chapter 2: Getting Your Goats

So, you've reached the exciting decision that you are going to buy goats. They will definitely transform your life for the better. All you need to do now is find some goats for sale, decide what breed you want, and then figure out how to transport them. This chapter will give you information about where to buy goats from, some of the obvious and unexpected costs involved when you have goats, how to spot a good goat, how to assess a goat's worth, and how to transport your goats home.

Where to Buy Goats

I would strongly recommend that you buy your goats from reputable breeders, who have a good name and reputation for their goats; rather than purchasing goats from auctions, especially if you're a beginner. Goats from auctions can be a great deal, but ultimately the goats won't have the same pedigree, or prior care that a good breeder will have given them. As a result, they may end up costing you more in terms of dealing with illnesses and disease, and therefore vet's bills, than what you at first thought you had "saved" by going down the auction route. But don't get me wrong, buying goats at an auction is a viable option, you simply have to be prepared and know what to look for.

There are multiple benefits to buying goats at an auction. It's a zero-pressure buying experience in the sense that nobody will try to get you to take home a goat. You can come and browse, check what goats are selling for and just do some window shopping. In addition to that, auctions often have a much wider selection in terms of breed variety and the number of animals for sale. Breeders and barn sales usually have limited selection of breeds and little diversity in the selection.

Here are some tips for buying your goats at auctions:

1. **Prepare**

Speaking of window shopping, that's exactly what I'd recommend to beginners. You have to go prepared, and you can prepare by being a spectator before becoming a customer. Attend a couple of auctions before buying your first goats if you decide to go the auction route. Take a seat and watch the action. This way you can learn how things work, get acquainted with the market and make appropriate bids and get familiar with the auction language.

2. **Ask for help**

When you're at an auction, don't be afraid to talk to the owner or operator, as well as the auction staff. They should be happy to familiarize new customers with the process and explain how you can register, how to bid, how you can pay and how you can take your goats home.

3. **A cheap goat doesn't mean a great deal**

Remember, the cheapest animals are often not the best deals, while the highest-priced ones are not necessarily the most expensive. It might seem like a confusing statement, but what it means is that the cheapest goats can become expensive really quick. They often have health problems and sometimes unscrupulous people sell their diseased goats at auctions. As a result, a cheap goat can often turn into a bunch of vets' bills. On the contrary, more expensive goats can prove to be worth every penny, so sometimes a more expensive goat is actually a better deal than a cheap one.

4. **Call ahead and arrive early**

If you're looking for a specific breed of goat, simply call ahead and the auction staff will help you with breed availability and other information. This can save you a wasted trip. If you go to an auction, then try to arrive early. This will allow you to check in the office, preview the goats for sale and see what animals will be auctioned.

5. **Take an experienced friend or colleague with you**

If you know someone who has some experience with livestock auctions, take them along if possible. They will be able to share their experience with you and you'll learn the ropes much quicker if you have someone experienced at your side.

How Auctions Typically Work

First of all, don't let the auctioneer or anyone else intimidate you. The process is actually pretty simple. Yes, there will be a lot of fast talking and prices flying around, but there's nothing to fear.

Check in when you arrive at the auction. You'll be given a number to be used for bidding. When you want to bid, simply raise your card. Keep in mind that at some auctions people can raise their hands for bidding, so avoid waving at someone or running fingers through your hair because you can end up with an unwanted animal. My advice is to never be the first one to bid. Auctioneers can often ask for more than they think they can get and then back up when they get no bids. Let someone else start and then jump in if you like the animal and the price.

Needless to say, if you're going to an auction not just to browse, but with the intention of purchasing goats, you'll need to arrange a way to transport your goats home, which will be discussed in detail later in this chapter.

There are some more places you can purchase goats from. If you are in the United States and wanting to find out where certain breeds of goat are located, and how to purchase them, I would suggest you contact your state's Cheese Guild for this information, and also try to tour creameries (places where milk/cream are processed and where butter and cheese are produced).

You can sometimes purchase goats from places that are visitor attractions, such as farm parks and petting zoos. Again, chances are that these animals have usually been very well looked after.

If you don't know any farmers or homesteaders who sell goats or can't find one, you can always check Craigslist. As mentioned previously, goats sold at auctions are not always the heathiest ones, and the same is true for Craigslist. You may have to look for a while, but you can certainly find decent goats for sale there, you just have to know what to look for and what questions to ask. Below, you will find all you need to know about spotting a good goat and assessing a goat's worth.

Spotting a Good Goat

You can assess the physical condition of a goat when buying it using a number of techniques:

1. **Take the goat's temperature**

You can use a rectal thermometer for this, a healthy goat should have a temperature of 103°F–104°F. You can hold a small kid across your lap, but if it's an adult goat, you'll need a helper to hold them or you can tie them to a pole or a fence. Lubricate your thermometer with petroleum jelly and insert it a few inches into the goat's rectum. Hold it in place for at least two minutes. Slowly remove the thermometer and read the temperature. Finally, wipe it with an alcohol wipe or a cotton ball that has been soaked in alcohol.

2. **Take the pulse**

First, make sure the goat is calm and resting. You can find the goat's artery below and slightly inside their jaw with your fingers. Watch a clock and count the number of heartbeats in

15 seconds. Multiply it by 4 and you'll get the pulse rate. The normal pulse for an adult goat is 70–90 beats per minute; however, it can be twice that in a kid.

3. **Eyelid color**

If you roll their lower eyelid down, it should be nice and pink. If it's pale, this could mean the goat has anemia.

Here are the questions you should ask the seller:

- When were the kids born?
- Who sired them? That's especially important if the buck doesn't live on the property.
- Ask about the history of diseases, worms and hoof rot on the property. Harmful diseases and organisms that live in the soil can be transferred between properties.

Goats are typically affectionate with humans. When you interact with the kids, see if they are frightened or become aggressive. This is usually a sign they were mistreated or neglected. A goat's tail usually points upwards unless it is frightened or sick. So, you can look for this as a sign of healthy goats, and to spot sick ones. Take a look around the property, check the goat shelter and see if it's clean, especially the bedding area.

Depending on where you live and what goat breed you're looking for, it can take weeks or even months to find decent goats. Buying goats can be a tedious process, but it's not as hard as it might seem, and it's definitely worth it.

I would strongly advise that you have the goats tested for diseases before you get them home, and then once again, once you have them home. Typical diseases for goats include Johne's disease, Caprine Arthritis Encephalitis (CAE), Tuberculosis (TB), Brucellosis and Caseous lymphadenitis (CL). These and other diseases will be discussed later in the book in Chapter 7, which is about goat health. You need to test goats before you bring them home so that they don't pass on diseases to your other goats and make them all ill or wipe out herds. If you get a vet to test your goat, this can be around $150. But, if you're able to draw blood yourself and send the samples off to a lab, this would cost around $20–$25 to test for Johne's, CAE, and CL.

Domestic goats come in approximately 200 different breeds and their sizes vary greatly. A Nigerian dwarf goat, for example, is a goat that produces lovely milk and it typically weighs around 20 lb, which is approximately 9kg. Pygmy goats typically weigh around 53–86 lb, or about 24–39 kg. Whereas an Anglo-Nubian goat can get up to 250 lb, which is approximately 113 kg.

Getting Your Goats Home

If you are just bringing a few goats home, you can fit several goats in a pickup truck, so you wouldn't necessarily need something like a horse trailer.

Goats don't really require special sized trailers; they can actually be transported in dog crates. You can fit 2–3 Nigerian dwarf or pygmies into a large crate; whereas if you have regular sized goat, you may require an extra-large crate. If you are purchasing adult goats, then you may need a large or extra-large crate for one. A buck old enough to breed should have his own crate. If you are transporting goats in winter it should be fine, but definitely give thought to how hot the goats will be, if you are transporting them in summer, and ensure it is safe, and a pleasant temperature for them. Be aware that goats may squat and urinate, so if they are in a vehicle, ensure that you have plenty of sheets down and waterproofing. You can also put things to absorb urine in the crate like straw or wood shavings. If it's a short trip of no more than an hour, this is fine. But if it's an hour or more, ensure the goats have hay to eat on their journey; and make stops so they can drink water. If you put water in the crate, then there's a good chance it would be knocked over.

Assessing a Goat's Worth

You may want to consider a goat's lifespan when you're working out the price of goats, and what you are likely to get back from them. Goats typically live from 8 to 20 years, but it's Boer does that can live up to being 20 years old. Most other breeds of goats live up to 10–12 years. Miniature goats typically live 12–15 years.

Dairy

Depending on what you want the goat for, can determine which goat you buy. Saanen goats are really good for dairy, producing 1–3 gallons of milk per day, but its butterfat is low, so it's not so good for cheesemaking. If you need milk with higher buttermilk content, an Anglo-Nubian goat will produce this for you. If you don't have a lot of land, then a Nigerian Dwarf goat is half the size of a typical goat and will produce a few pints of milk per day.

Meat

If you are planning on having goats to produce meat, then I have some good news for you. As mentioned previously, goat meat prices are on the rise at the moment, in 2021 a pound of goat meat typically cost $3.50 to $3.80, up from $2.50–$2.75 5-year average. However, keep in mind that prices always change and you should always do market research before starting a business.

30 million dollars worth of goat meat is imported into the US each year, and over half of this is from Australia. As a result, there is a vast market for locally produced goat meat.

Savanna and Boer goats are good as meat goats because they grow quickly and have year-round heat cycles (which means they can breed year round). They are quite expensive, however. Whilst a Boer doe can cost $600, and a buckling Boer can be $1,200, these still tend to be a good investment

Fiber

When you shear an Angora goat, this is likely to yield about 5 lb of mohair. Typically, you would shear this type of goat twice a year.

With cashmere goats, the quality of cashmere fleece depends on how long it is, its diameter, and how crimped it is. Cashmere fiber is crimped, soft and should lack sheen. It needs to be at least 28–42mm long. It needs to have a diameter of less than 19 microns. Cashmere wool is usually 7–8 times warmer than sheep's wool, yet is lighter and softer.

Approximate Costs of Starting a Goat Herd

Whatever your reasons for wanting goats, whether that's for milk, cheese, meat, or fiber, there are costs involved. Some are obvious; others not so much, that until you've had the experience, you wouldn't realize the hidden costs. This section is here to give you a much more thorough understanding of the approximate costs of starting a goat herd. You will need to do some research and budgeting before starting to purchase goats. Always give yourself financial leeway to account for the unexpected things, such as veterinary bills and equipment breakages.

Cost of Goats

Goats themselves really vary in price and can be a few dollars per goat, up to hundreds of dollars and this can be determined by whether they are male, female, a buck or wether (neutered), whether they have horns or not, their age, breed, whether the goat is registered, its health, and where you purchase them from.

Goats are definitely cheaper than some other animals, such as horses. Goats cost twenty percent of what horses or mules cost. They don't require too much space, so if you have a small area, goats are a great choice.

If you want bucks ready for breeding goats, you can look for bucklings. These are young male goats that have not been castrated. When you purchase them, they will not yet be ready to breed with, but this can be a way of buying bucks without a lot of expense, from places like dairy farms. If you buy goats from a good reputable breeder, they will be more expensive than from a sale barn, but this should save you money in the long term if you want to produce milk or breed, because this is a guarantee of good health.

If there's a surge in goats as pets, this may cause the price of them to increase, as it's been the case with Pygmy goats. If you're wanting to buy goats to eat weeds, then any breed of goat will be sufficient to do this. Buying wethers for this task can be sensible, as they are friendly and won't be pregnant or injure udders whilst weeding. If you want a pet goat, bucklings from breeders may be quite inexpensive. If you want pygmy goats, these can be just as expensive as full-sized goats, if it is a purebred miniature goat, such as a Nigerian Dwarf or Pygmy.

Milking Equipment

Chapter 9, later in this book, is all about milking. So, there's just a brief overview here, just to give you an indication of costs. But the later chapter will go into the process and procedure of milking more thoroughly.

If you are purchasing goats specifically for the intention of milking, then you would need a doe who has been registered, dehorned, and disease tested, and the cost of this is typically between $250 and $500 and depending on the breed sometimes up to $1,000. Dairy goats from breeders have been specifically bred to give you the most milk, from the least amount of feed, but whilst ensuring the animal has enough and is healthy.

If you have a milking stand, this costs between $100 and $500 depending on the quality of it. If you are hand milking your goats, you will need a pail, teat wipes, mastitis spray, strip cup, filters, and a funnel, which will cost together approximately $150. If you decide to use a milking machine, that can cost between $500 and $1,500.

Veterinary Costs

When you are more experienced at looking after goats, anything you can do yourself to save spending money on vet costs is preferable, such as testing goats, buying thiamine, and being aware of the symptoms of goat bloat. But, for a beginner, to keep your goats in the best health, it is good to have a reputable vet on hand. There are various vaccines you can give to goats on a schedule throughout the year, for CDT and pneumonia.

Whatever you think you will spend on vets' fees, my advice would be to double or even triple this, just so that you are definitely covered to take good care of your goats. If you don't need it—great; but it's nice to ensure that the health of your goats is a top priority. I would advise to take out insurance to cover vet costs too, which won't likely cover day-to-day maintenance of hooves and worming medication, but would help if you have a seriously ill goat.

If you feel confident about trimming your goats' hooves, then trimmers will cost approximately $30. You may need to sharpen them from time to time too (sometimes farm shops where you buy straw or feed from may have a sharpening service). If you don't want to trim hooves, then it will cost you around $25 every few months to have this done. Are you happy to worm your goats yourself? If not, this will usually cost approximately $100 per goat twice a year, for fecal samples, vet fee and wormer medicine. If you simply have a couple of goats, this is manageable. If you have herds of goats, this is a larger chunk of money to consider.

If you want to have your goats castrated, this usually costs around $100 at the vets. You can do it yourself using goat castration bands plus CDT vaccines to prevent tetanus.

Breeding Costs

Chapter 8 later in this book will give you more detailed information about breeding, but this is simply to consider costs involved. If you want to breed goats, there are a number of options available to you. You can rent a buck, buy a buck, or use Artificial Insemination (AI) or there is driveway breeding. If you rent, this can cost $100 per doe, plus board and feed. If you decide to buy, it is whatever the buck costs plus housing, fencing and feed. AI can cost hundreds of dollars per doe plus shipping, and the help of an AI specialist. Driveway breeding is where you would take your doe to a buck whilst the doe is in heat for a short visit and hope that the doe was impregnated, this can cost $100 per doe.

Food Costs

Grain for goats can be approximately $12 per 50 lb bag. You can buy Alfalfa mix for approximately $10 per bale. A ton of Alfalfa hay costs around $165. For optimum health, goats like quality hay, fresh vegetables, and you can supplement a goat's diet with selenium paste, other minerals, vitamin B, and treats, which are likely to cost $50 per month.

A goat should be eating 2% of its body weight in hay per day, unless it's pregnant, lactating or working and it will eat 4%. So, a 100 lb bale of hay should typically last 45 days for a 100 lb goat, or 22 days if the goat is pregnant. Through winter months, goats need more hay to keep warm, especially if there aren't other plants for them to forage in the winter. Some people have a feeding schedule, others let the goats eat freely when they wish to. We'll look more at feeding goats in Chapter 5 of this book.

Not to sound doom and gloom about food for goats, but do ensure that it is stored safely away from flooding, and without mold impacting it because this could mean the food is wasted. Never feed moldy hay to a goat because this could cause listeriosis.

You can buy sweet feed for pregnant, lactating or poorly does, but don't feed this to wethers (they should only be eating hay and what they forage).

Other food for goats can include what people eat, but this should be as a treat, because goats do need specific minerals including phosphorous.

You can buy goat minerals to ensure your goats are not deficient in copper or selenium, and this can cost $20 for an 8 lb bag, to over $100 for 50 lb, this will depend on the brand you buy too.

Initial Setup

When you first start looking after goats there are lots of other costs that you'll need to consider as part of setting up your property to house goats, the next two chapters of this book look at housing, bedding, and fencing, which you'll need to account for when you are putting together a budget for goat costs.

The first two goats that I ever bought were Alpine goats called Emma and Ophelia and I bought them from a good reputable breeder, who I've continued to buy future goats from over the years because their goats are well cared for and healthy. I'd rather pay a bit more initially and avoid ill diseased goats from a non-trusted seller. Emma and Ophelia are fantastic milk goats, and I still remember the very first time we had home produced milk from them, and later homemade goat cheese which was delicious and creamy and seemed to taste much better than store-bought cheese. It's an incredible feeling of caring for the goats and getting wonderful produce in return that can make you more self-sufficient. They are absolutely worth every penny I've invested in them.

Key takeaways from this chapter:
1. It's best to buy goats from reputable breeders.
2. The costs of starting a goat herd are likely higher than you anticipated, this isn't anything that should put you off, but do consider the hidden costs too and account for more than you expect.

3. You can test goats before you buy them to ensure they are free of disease, and there are signs you should look out for.
4. Buying goats is an investment, which can be hugely profitable and change your life, making it much more sustainable.
5. It's relatively easy to transport goats without any special equipment, they can be moved in dog crates when the goats are young.

The next chapter will move beyond transporting the goats to your home, to looking at how best to create comfortable, safe, and essential housing for your goats.

Chapter 3: Housing Your Goats

Whilst goats are sturdy creatures, they definitely need adequate shelter and housing to keep them protected from the elements. Some goats dislike getting wet and will want to shelter in a barn, and others deliberately avoid puddles that would make them wet and dirty. They need protection from the cold and rain, as well as the extremes of weather, such as snow or scorching summer heat. You will definitely need to ensure that you have adequate tight fencing, and the next chapter will cover fencing in much greater detail. This chapter, however, will focus on building a shelter for your goats and knowing how large it needs to be, the bedding you require, a manger/trough from which your goats will eat, helpful additions, and security considerations.

Goat Shelter Considerations

Shelter for goats doesn't need to be really extensive or complicated. A good shelter needs to be a good size, clean, dry, ventilated, and it needs to provide protection from the elements. Let's take a look at all these factors one by one.

Size Requirements

A three-sided shed or a barn which is large enough to protect the goats from wind, rain, and snow should be sufficient. A number of goats are able to share a shelter, but if a doe is pregnant or has kids, she will need her own space. If your goats are getting into scuffles over food or sleeping space, you may need to separate them, for example putting them in separate barns.

If your goats during the day can access a lot of space in the way of woods and pastures during the day, then their space to sleep in can be approximately 15–20 square feet per goat. If they don't have a lot of space outdoors throughout the day, then they should have 20 square feet per goat, and at least 30 square feet area outdoors to exercise in—these are the bare minimum requirements. If you have pregnant goats, they ideally need a space of 4 by 5 feet at a minimum, so if you are planning on breeding goats, the size of your shelter will determine how many does can breed at once.

Goats love things to jump on to keep them active and entertained, things like levels, shelves, straw bales, picnic tables, and custom–built platforms are all excellent things to keep goats occupied and happy, do ensure they're positioned not too close to fences though that they could then jump over.

Goats obviously vary in size, and the bigger the goat, the more space they need. If you have a small herd of just 2–3 Nigerian Dwarf goats, you won't need a full-sized barn. But of course, if you had 30 of them, well then you need a bigger area. As previously mentioned, pregnant or nursing goats do need their own space with their kids if they're nursing. Do keep in mind that goats will grow, so whilst you may not need as much space for kids, they are going to grow into full sized goats, and you need to account for that and ensure you have the space to house them comfortably.

Regarding adequate pasture size, ideally this is one acre of land for every 10–12 that you own. You need to ensure that the area has enough shade too so that animals aren't constantly in the sun. If you don't have as much space for your goats as you'd like, it can be a good idea to take your goats for a walk on a leash, they can be trained to do this.

Manure Management

You will have to keep your goat shelter clean, so you have to make sure it's human-accessible. Having to crawl under a low roof can make maintaining your goat shelter quite a challenge. Being able to move freely is essential for maintaining your goat shelter, replacing the bedding, or adding fresh layers. So, if you're building your goat shelter from scratch, keep in mind you have to make it human-accessible too.

Dry Floors

Dry floors are essential for goats' health. There are a few options, such as dirt floors, wooden floors, or even expensive concrete foundations. All of them have their pros and cons, of course.

Concrete floors are really solid; however, they require daily spray cleaning. Obviously, you need to have a pressurized water system for that, which can be hard if you live off-grid.

Some people recommend dirt floors as a preference over wooden floors, because wooden floors can sometimes get slippery, and you don't want a goat to tear a ligament or do any joint damage on a slippery floor.

Dirt floors may "breathe" more naturally, but they are also can get directly influenced by rainstorms. You can mitigate these effects by building your goat shelter on sloped ground so that moisture can drain downhill instead of pooling in the shelter.

When you're cleaning out goat enclosures and have removed any poop and replaced damp straw, then you can also put hydrated lime on damp areas, which will dry these up, and help to reduce bacteria.

Bedding

There are a lot of different options for bedding; however, typically goat owners tend to use wood shavings or straw for bedding in the goat's barn. You can also use softwood pellets (pelleted bedding), cedar chips/shavings, sawdust and even sand. But in most cases, it really comes down to pine vs straw.

Pine shavings are typically used as primary goat bedding material. They are really absorbent, reasonably priced, and easy to manage. It's easy to spot clean wet areas and spent pine shavings can be added right to the compost pile.

Straw isn't the best choice for primary goat bedding. It can be cheaper and some people find it less messy when dry. It doesn't absorb moisture as well as pine shavings; however, and although it's easier to clean out with a pitchfork, it can get messy because it doesn't absorb pee as well as wood shavings, so it can sink down to the ground the floor. However, straw holds the body heat really well, so it's a great option for the winter time.

In the end, it comes down to what works best for you and your animals. You can try both options and see which one works best for you.

Goats pee a lot, so I highly recommend using stall freshener regularly. The ammonia contained in their urine can become toxic to their lungs. Stall freshener comes in different forms, but I prefer it in the powdered form. Simply sprinkle it all over the floor of your goat shelter and add a little extra when goats seem to pee a lot.

In addition to that, you can use lime wash or lime powder to keep everything clean. Stall freshener is usually more effective, but barn lime works well too. Lime wash and lime powder are two different things. Lime powder is used on dirt, it strengthens soil. You can use it in your garden too. Lime wash is used like a paint to whitewash surfaces. It not only paints the surface white but also has antibacterial properties, and it's often used in chicken coops. We used to paint the inside of our goat barn with interior paint, but a couple of years ago we decided to just give it a coat of lime wash, and it's been great. It's cheaper than using paint and more environmentally friendly, plus it has antibacterial properties which is great.

Now, let's talk about cleaning your goat shelter, what supplies you'll need, and deep litter vs non-deep litter method. Here's what you need to clean your goat shelter:

- A wide shovel (snow shovels work great too)
- A large wheelbarrow
- Stall freshener
- Fresh bedding (we use pine shavings)

Deep Litter Method

Deep litter method is typically used in winter to create additional warmth for your goats. It's essentially a giant compost floor. You shouldn't use deep litter method during the warmer months because of flies. But in the beginning of fall, as the temperatures start to drop, you can begin your deep litter.

Get everything out of your goat shelter, including your goats, of course. Deep clean it, you can use a diluted bleach solution, the rinse, and allow it to dry completely. Sprinkle a generous amount of stall freshener all over the floor. Now layer straw or pine shavings over it. As mentioned previously, straw holds the heat much better, so if gets really cold you should use straw as bedding material. When the floor becomes noticeably dirty, simply layer fresh bedding on top without cleaning. Just keep covering it up all winter long. The manure and urine will break down and the bedding will provide additional warmth in the winter months.

Non-Deep Litter Method

With the non-deep litter method, you'll simply have to clean your goat shelter regularly. We usually clean our goat shelter every 7–10 days, and it typically takes about 30 minutes.

Cleaning Your Goat Shelter

To clean your goat shelter, bring in the wheelbarrow and start scooping out the bedding. You can use a broom to get into hard to reach areas. Once you clean everything out, sprinkle some stall freshener all over the floor. Then, layer fresh bedding on top.

Protection from Extreme Weather

It is advisable to find a dry and higher place for the goats so that if there is any flooding, the goats will remain safe. The floor of the barn should be dry. The place should be light and airy, so you have the ability to control the temperature and moisture as best you can. You don't want the barn to have any damp, because this can cause diseases in your goats. You don't want rainwater to be able to enter the barn. It needs to be strong, comfortable, and safe, so your goats can happily rest. In the winter months, you may need to use a water heater so that the goats' water doesn't ice over.

A three-sided is sufficient or a pole barn is sufficient, provided that they have some protection from the elements. Goats do need fresh air, but the area they sleep in needs to be draft free, because this can make goats ill and give them pneumonia and scours. Greenhouse barns, calf hutches and large dog boxes can provide shelter for goats.

Now, let's talk about temperature and protection from the elements. Goats will try to get away from bad weather as soon as they can, unlike sheep, for example. Consequently, you'll have to make sure they can get out of the wind, rain, and snow whenever they need. Goats are actually pretty sturdy creatures, in the winter they can keep each other warm all the way down to 0 degrees Fahrenheit, but only when they are dry and there are no drafts that can suck away their communal heat.

You can check if your shelter if wind-proof by pretending to be a goat. It may sound silly, but it's a great way to check if your shelter has any drafts. Simply hunch down to their level and see if you can feel a breeze. If you can feel it, then your goats will feel it too. Do your best to find gaps that make the breeze and seal them. However, don't overdo it and don't make your shelter so airtight that air can't move at all. Good ventilation is a must to keep your goats healthy. I highly suggest making sure there is some ventilation somewhere around the perimeter of the roof area. That way, there will be good ventilation and the air will be able to move freely without blasting your goats directly.

Goat shelter is equally important in the summer as it is in the colder months, as it will provide protection from rain, wind, and heat. While some desert breeds can actually appreciate a hot sunny day, some other breeds that originated in colder climates don't like the heat so much. If the temperatures rise above 80 degrees Fahrenheit, they will certainly seek shade for relief.

Creating a cross breeze in the shelter can help keep the air fresh and provide additional cooling. You can build a removable panel on the wall opposite the door. In the summer, you can remove it and cover the hole with a goat panel to allow the air transfer and keep the goats in. And in the winter, you can simply put the panel back and seal it to prevent winter drafts.

You can build goat houses on the ground with bricks and cement and put dry straw on the floor. Or you can make goat houses with poles that are approximately 5 feet off the ground, and this can keep the goat safe from flood waters or damp conditions.

A concrete style house is more expensive, but they're easy to clean and they will ensure the safety of your goats from predators.

If you have goats for milk and will be looking after baby goats all year round, then I would definitely suggest that you invest in goat housing that will last for years, rather than anything more temporary and makeshift.

You can buy pre-packaged kits for goat shelters and enclosures that are almost tent-like in their construction. Keep in mind that goats chew on things, so ensure your shelter is constructed out of things they can't destroy.

The Manger

A manger is a trough from which your goats will feed. Some barns are constructed to have these almost like guttering on the outside of the barn so that the goats can put their heads through and eat.

In the picture above, you can see the barn has a corrugated iron roof, which will deflect the rain off and keep the goats dry, it's raised on stilts which would protect the goats from flood water or damp, and there is a manger/trough guttering on the outside of the barn. It has wooden flooring, which is very easy to clean, it's light and airy and less prone to disease. It doesn't have a lot of protection from the wind and cold, however, so it's won't really work in the cold or windy areas.

Your goats will need hay in the barn or manger. Even if your goats live in a pasture all the time, they will still need hay and not just grass to graze on. It is best to have a hay feeder because goats don't like to eat off the ground, you will waste a lot of hay if it's on the ground and it can be prone to parasites. It is worth investing in a proper one, rather than attempting to make a

---

[5] Image from Roys Farm: https://www.roysfarm.com/goat-housing/

makeshift version. It is important to have good quality hay. They may want alfalfa, and things like tree bark and forage to give them fiber.

If you are giving your goats grain, they will also require a grain feeder that can be easily accessed. If grain falls onto the ground, the goats will not each it from there so that is wasted grain at that point. You will need a mineral feeder too.

Goats don't always use things in their space in the way that you anticipate they will. A goat manger with a slipped lid may be climbed upon (and get broken), other mangers may be laid in by the goats, then they will feel the hay isn't fresh to eat.

You will also need food storage containers to keep the food dry and away from the goats themselves, rodents, birds, or other animals. Food storage will be discussed a bit later in this chapter.

Additional Facilities for Breeding and Milking

If your goats are kidding (pregnant or lactating) then they will need a more solid building to keep them sheltered. Within the building you can get livestock panels to separate the space into pens for each doe with her kids. This separate safe and clean area will provide the does and their kids with a private environment that they need and will also allow you to monitor them during the first few weeks. Does and their kids are the most vulnerable to cold. If it gets really cold, you can consider installing a heating element over the kidding area, but keep in mind it creates a fire risk that has to be taken into account. Alternatively, you can time the breeding so that does only give birth during the warmer months.

And the final touch to make your goat shelter truly multifunctional is building a milking area. This will make milking easier, as your does will get acquainted with the milk stand and you won't have to move them to different spots all the time. Keep in mind you'll have to make the milking area inaccessible to goats when you're not there. A 4-foot-high barrier should be enough to prevent them from roaming around the milking area when it's not in use.

Helpful Additions

When building a shelter for goats, it is worth considering where you will store their food (hay/grain/minerals) and bedding (such as straw or wood chips) plus any other goat equipment you regularly use. You'll have to build a 4-foot-high barrier to keep your goats away from this storage space.

You will also have to prevent bar rodents from messing with your goat supplies. You can store grain in galvanized trash cans. It may sound a bit weird, but the lids make a nice and tight seal, and there is no way mice can chew through metal.

Finally, you can consider building a composting area for the manure and soiled bedding halfway between your goat shelter and garden. It's a great storage area for garden fertility when you're in between seasons.

Safety Considerations

The most important consideration to keep in mind is the placement of your shelter in regards to your perimeter fencing. Even if you have a really strong and secure fence, if you place the shelter next to the fence, you will provide your goats with an escape route.

Make sure your goats can't climb on the roof and take a flying leap. Even if they can't reach the fence, keeping goats off the roof is a good idea in general. Goats can damage the roof with their sharp hooves, and even though they are really agile animals, they still can fall off the roof and break their legs.

Young goats can be extra cheeky. They enjoy leaping against walls and pushing off with all fours. That's the reason why there shouldn't be any glass in your goat shelter. While a window may look cute, the goats will eventually break it and hurt themselves.

My final advice is to inspect your goat shelter often and check for signs of wear and tear and any potential hazards. Goats love to mess around with structures, fences and pretty much everything they can find, so they can create their own danger occasionally. Keep an eye out for any signs of wear and tear, such as boards or nails sticking out or pieces of stray wire, and fix them as soon as you can. Otherwise, it's just an accident waiting to happen.

Our goats, and certainly every other goat owner I've spoken to over my career, have said that their goats like to keep out of bad weather, if it's raining, windy or cold! I don't blame them, I do too! Not one owner I've met yet has ever suggested that their goat likes the rain and wind. They do love the sun and a nice fresh breeze.

When we bought our first two goats it was summer and we had a pallet shelter, but as winter approached, we moved them into a much studier concrete barn, which keeps them warm and protected. They have the choice of where they want to go and they do spend a lot of their time outside, unless it is windy, rainy, or snowing. I now have a wide collection of goat shelters for them to pick from; some are hardier than others, and some I've constructed out of wooden pallets, posts, and corrugated iron roofs. There are dog houses in the grounds that they use, pole barns, regular barns, and other options. If you're handy with DIY and have building materials left over from your other projects, you'll certainly be able to build a range of goat shelters.

Key takeaways from this chapter:

1. Goats definitely need good, safe, draft free housing that has some storage room for their hay, feed, and other goat supplies.
2. They need lots of access to good clean water, minerals, and quality hay.
3. Goats need at least 15–20 square feet of space per goat. You would benefit from safe indoor individual spaces for pregnant or nursing does and their kids.
4. There are a really wide range of different shelters you can build for goats, from more temporary makeshift to permanent, but if you have goats for milk, it's a sensible investment to look for permanent housing.
5. Bedding can be made most typically from straw or wood shavings.
6. You will need a manger, troughs or a hay-feeder for your goats.

If you get the above things set up well, these will help with your daily routine of looking after your goats, and getting into a routine of cleaning, changing water, feeding, and topping up loose minerals.

Having good housing for your goats can help them to stay in good health, free from draught related diseases.

The next chapter looks at the fencing that you will need to keep your goats secure within a perimeter, and interior fencing needs too, that can separate goat groups.

Chapter 4: Fencing

When we first got our goats, we'd obviously done a lot of research and had tried to anticipate our goat fencing needs, but if I'm being truthful, a lot of our fencing was trial and error, and if our goats escaped, we then tried again by reinforcing the perimeter fencing. This chapter is here to give you the benefit of our experience so that hopefully you can be really well-informed and get things right the first time.

Goats are naturally intelligent and inquisitive creatures; you don't want them escaping from your enclosed area, because you may want to keep them away from flower beds, vegetable patches, neighbors' gardens, or dangerous roads. Goats jump very well, so your fencing needs to be high enough to prevent them hurdling the fence. It's not at all unheard of for people to install 4-foot-high fencing, thinking they've done a fantastic job, and then shortly after see one of their goats effortlessly jump straight over the fence. They will test your fence boundaries, and if there are any gaps, they will find them and get out!

Goats are known for climbing, balancing, and being very well coordinated, they will also work together climbing on one another's backs. Goats can climb trees, depending on the angle of the tree. This chapter is here to ensure that you provide suitable fencing to keep your goats safe at all times. It has been mentioned previously that if you are buying goats as pets, it's kinder to the goat to buy more than one, goats are very social and don't like to be on their own. If you bought one, it is more likely to escape or injure itself trying to escape the fencing to get to you. If you have more goats, they are happier together and won't try to escape as frequently. Myotonic Goats are not goats that tend to clamber or fence jump, so these could be ideal goats when considering fencing.

Fences keep your goats safe and secure, but they can also be a deterrent to predators, such as coyotes, foxes, dogs, and bears, depending on where you live. You will need fencing around the entire area that you intend to enclose your goats in, but then, within that space, you may want to have other fencing to have goats separated from one another.

This chapter will look at what kind of fencing you will require looking at interior and exterior fencing. It will cover gates, latches, fencing, and investigate the amount of fencing you require to keep your goats in a certain area of your property, keep predators out, and to separate goat groups from one another at certain times.

Goat Fencing Considerations

Here are the basic considerations you should keep in mind when choosing and installing fencing for your goats:

Area

In my experience, about 250 square feet of outdoor space is enough for a goat. As mentioned previously, you shouldn't keep just one goat, so you would need at least 500 square feet of outdoor space for a couple of goats. Ideally, you should keep up to 12 goats per acre. Naturally, the more space your goats have, the happier they will be.

Height

Most people usually recommend making 4-foot-high fencing for goats; however, I would suggest increasing the height to 5 feet. Goats are very active animals and they often can jump over a fence that's 4 feet high. Also, keep in mind you shouldn't leave gaps along the bottom. Goats can crawl in unexpected places and they can certainly crawl under fences if a gap is big enough. This is especially important if you have small kids. They are eager to explore the world and will try to crawl everywhere, so make sure your fence doesn't have a big gap at the bottom.

Head Gaps

Goats are curious animals and they love to shove their heads between things. While it might look adorable, it can be a deadly mistake if they have horns. Make sure that any gaps in your fence are no larger than 4 by 4 inches, be it squares of a wire panel fence, spaces between posts, or cross braces.

Obstacles

Goats are fun-loving animals, so providing them with various obstacles, raised platforms, toys, logs or even construction spools is a great idea. Watching your goats leap, prance, and balance is a joy; however, keep in mind that every raised surface should be at least 5 feet away from the fence. Pay attention to low-hanging tree branches as well, as goats can use them to make a running leap and clear the fence.

Attachment of the Fence

I would recommend attaching the wire panels to the inner surface of the post instead of the outer surface. Goats will try to push the fence, and when the panels are attached to the inner surface of the fence post, they will be pushing the hardware into the post and not out of it. Gates should be hinged so that they open into the goat yard, and not outward. That way, even if your goats somehow manage to unlatch the gate, they will be pushing the gate closed as they lean against it, rather than pushing it open.

Walk Your Fence Line Often

I would suggest making a habit of walking your fence line so that you can check your fence often. Keep an eye out for potential problems, such as sagging, chewing, or gaps that can form from goats pushing against weak points. The best way to avoid an accident is to prevent it from happening in the first place.

Do be prepared that you may have to swiftly change your fencing plans if things haven't quite worked as first hoped. Keep checking all your fences to make sure it is all sturdy and there aren't any gaps that have appeared. Goats like to use wire fencing as a scratching post, but when they constantly scratch in the same spot, this can weaken the fencing and make it worn.

Types of Goat Fencing

Below, you will find all the necessary information on different types of goat fencing, including the pros and cons of each type of fencing. The options are organized in order of typical cost from cheapest to most expensive, but keep in mind it's really hard to give hard numbers.

Material cost differ from store to store, and if you don't install the fence yourself, installation cost can vary greatly. Not to mention that the area you need to fence will increase the cost exponentially. With that said, let's take a look at the different options of goat fencing.

Wooden Fence

Wooden fences can be built from materials you have on your homestead and generally they are the cheapest to build. Installing the wooden fence does require a lot of hard work, though. Driving posts into the ground is not the easiest job in the world and you will have to maintain a wooden fence constantly. In addition to that, you will need to use a lot of material.

If you decide to go with the wooden fence, I'd suggest building the fence stockade-style, and not picket-style in a buck's area, as goat hooves can get trapped when they stand on their hind legs to look over the fence.

Pros and Cons of a Wooden Fence

Pros

- It's the cheapest to build.
- Unlike an electric fence, you don't have to worry if it's working. As long as it's standing, it's "working" and it's relatively cheap and easy to maintain.

Cons

- Goats can chew on wood and weaken the posts; they are great at exploiting any weak point they can find.
- Wooden fences tend to weather and rot with time, they require regular maintenance.

Wooden Fence Cost

Wooden fences are potentially low cost, especially if you mill the timber yourself, or if you already have an existing fence. Getting a service to pound posts into the ground for you will increase the price quite a bit, but it's still typically the cheapest option.

Electric Fence

Electrified fencing hedges in animals using a psychological rather than a physical barrier. It's long lasting, and it's very easy to use. If your farm area doesn't have easy access to electricity, this isn't an issue, and you can buy electric solar boxes that you place on the fence, with a steel rod going into the earth to ground it. Unlike many types of livestock, you'll need to put the fence's charge higher than you may expect for goats—somewhere from 4,500 to 9,000 volts at all times.

Goats are smart animals, and if they know there's a time when the fence is off, they will figure out how to use that to their advantage. Even if your fence has a really high charge, you may still find it might not be enough to stop a stubborn animal. Many goat farms tend to use high-tensile wire in combination with electric fence to keep their goats safe. You can purchase high tensile wire fencing for perimeters, typically you would pay someone to install this for you and ensure that it is working correctly, unless you have knowledge in this field. They are effective, and good for places that have harsh winters. They are permanent and secure, but more expensive. I would personally suggest that if you are just starting out with goats, you pick a less expensive and less permanent option initially, but if you are moving past your second year of having goats and determined that the goat life is for you, then this is a sensible investment.

You can soon take up electric fence panels in winter months and store them until you want them again, if you fold it up like an accordion and put it around a piece of plastic, it can be kept in a barn or basement until needed. You don't have to package up your electric fencing in the winter, but try to remove snow from it if it snows heavily, to prevent the fencing being stretched and saggy by the weight of the snow. The only downside to this is if you have soil that is very dry, and therefore the posts don't stay in well, and also if your goats are very lightweight, then they won't receive much shock at all from the fence.

If you have electric fencing, you'll want to train your goats that it's there, ideally starting from when they are kids. You don't want them hurt by it, but a very brief shock to them when young will teach them not to put their heads or bodies near it. Goats are really smart animals, and in most cases, a zap or two will teach them to stay away from the fence. If you pick a time when the ground is wet, this will ground the goats. It is extremely important to be there and next to the power box when training kids, just in case a young kid panics when it is shocked and tries to carry on going through the fencing, rather than pulling its head away. You don't want it to get stuck and continue to be shocked, this would be horrendous. But if you're there for training and control the power, they will learn to respect the fence.

Do make sure you have selected the right amount of charge for the fence and ensure that if a human touched it accidentally, you are not going to cause any damage to them. Electric fencing is relatively inexpensive, and easy to put up. If you have an area that is mountainous or with lots of woods, woven wire can be difficult to use here, but using T-posts and an electric fence is much easier. The majority of goats won't try to pass an electric barrier, but they seem to have a sixth sense about when the electric is not working and will try to go through it then. Electric fencing is not always enough for bucks in rut, and you will need a physical barrier to stop them from going where they want.

If you have an electric fence around where your goats are, do ensure that you regularly keep the weeds and grass around it trimmed to keep it in good order because otherwise they will drain the battery power when they touch the fence and it won't work effectively. It's also worth checking that there are no fallen tree limbs that are touching it, so regularly walk around the fence area to check it is all good. Another important thing to look for when you are checking fencing are the holes that have been dug near the fencing by predators, if they dig a big enough hole they can get under and into your goat enclosure so look for these and fill them in and board them up accordingly.

Pros and Cons of an Electric Fence

Pros

- Easy setup.
- Relatively affordable.
- Easy to move if want to try a rotational grazing method or brush control in different areas.

Cons

- There are a lot of things that can short out an electric fence.
- Electric fences require training, goats need to learn to respect the fence.
- Weed control is extremely important in order to prevent tall weeds or grass from rendering your electric fence useless.

Electric Fence Cost

Electric fencing is a cheaper option if you want to try a rotational grazing method, but haven't been able to put a wooden perimeter fence in place yet. It requires a lot of maintenance to keep it running, as there are a lot of things that can short it out and weed control is extremely important if you have an electric fence, so consider that if you're looking into installing an electric fence to keep your goats safe.

Woven Goat Wire and Field Fence

Woven goat wire is a great option for permanent fencing, and it's the one that I usually recommend. Make sure you choose the goat-specific option with 4x4 inch holes, rather than the larger weaves, such as 6x6, 6x9, and 6x12, which are used for larger livestock. While it's more expensive as there's a lot more material used to make the denser weave, it will prevent horned goats from getting their heads stuck.

Field fence is similar to woven wire and it may also work with some caveats. Field fence is designed for horses and it's often made with a finer gauge wire. That usually makes it cheaper; however, it also stretches and bends out of shape more easily than woven wire. In addition to that, field fence has a much wider weave. Goats love to climb, and they can balance on really small surfaces, so they can sometimes get their heads stuck in a field fence.

Woven wire fence is relatively easy to install. You simply drive the steel posts into the ground every couple of feet, then attach the panel to the post with zip ties or wire clamps. However, make sure you install this fence nice and tight. I'd suggest cementing the posts in the ground. If the posts are on the outside of the fence, this will give you a more solid fence. Goats are likely to stand on it and push it, and when the panels are attached to the inner surface of the fence post, they will be pushing the hardware into the post and not out of it.

Panels that are sold specifically for goats are typically 4 feet high, however you can find 4x4 inch woven wire fence panels that are 5 feet high, so I tend to use them instead because these are a bit higher to prevent the goats from jumping over them. Alternatively, you can string a line of electric wire above the top of the fence or use it in combination with a higher, wooden frame. The panels will be sold either where you purchase your feed from, or available from many places online. This is a really good, sturdy, and trustworthy fencing option that I highly recommend, and it's the one that I personally prefer.

Pros and Cons of a Woven Goat Wire Fence

Pros

- Strong and dependable.
- Probably has the best balance between cost, ease of installation, reliability, and longevity.
- Dependable, strong, and one of the more often-recommended methods for fencing goats.

Cons

- Standard 4-foot-high panels can be short for some breeds; however, you can find 5-feet and even higher panels relatively easily. Alternatively, you can string a line of electric wire above the top of the fence or use it in combination with a higher, wooden frame.

Woven Goat Wire and Field Fence Cost

Woven wire panels are typically the middle of the road in terms of cost. They are not the most expensive option, but not the cheapest either. They are not very difficult to install, so you can install them yourself; or have them professionally installed if you're not too keen on doing it yourself. A 100-ft roll of 4x4 inch woven wire that is 5 ft high typically costs between $70 and $85. Some places have quantity discounts when you buy 10 or more rolls, for example.

Cattle Panels, Stock Panels, and Goat Panels

Cattle panels are solid metal panels and they are a good option for making a strong fence. They tend to be somewhat expensive, though. But even if they are out of your budget to use as fencing, I'd still highly suggest using them for sectioning your barn, especially during kidding season.

While these panels are as solid and bend-proof as you can get, they do have one drawback. Kids can easily escape from these "as is" because they have gaps at the bottom. This can be solved by adding some chicken wire or hardware along the bottom of each panel.

Pros and Cons of Cattle Panels/Stock Panels/Goat Panels

Pros

- These panels make a fantastic perimeter fence, they are incredibly solid and bend-proof.
- They don't rot and won't bend out of shape.

Cons

- This kind of fence can get quite expensive.
- Kids can escape as there is a gap at the bottom; however, it can be solved quite easily by adding something like chicken wire along the bottom of each panel.

Cattle Panels/Stock Panels/Goat Panels Cost

While cattle panels are strong, they do tend to get quite expensive. A 16-foot cattle panel usually costs around $20 apiece, and that doesn't include any of the wooden posts that would be used to install the panels. Furthermore, cattle panels usually have spaces that are designed for cows, not goats. They will require some modification to work. Specifically designed goat panels will cost around $60 apiece.

Chain Link Fence

Chain link fence is probably the most goat-proof option, but it is also typically the most expensive one. If you have a small herd and can afford a chain link fence, it is well worth considering as a long-term permanent solution. It is great at keeping the goats in and predators out. If you can't afford it as perimeter fencing, you may still consider it as an option for containing your bucks.

Pros and Cons of a Chain Link Fence

Pros

- Solid, sturdy, and long-lasting solution.
- Probably the most goat-proof option.
- Great at keeping out predators.

Cons

- Can get really expensive.

Chain Link Fence Cost

A 50-ft roll of 11.5-gauge 4-foot-high chain link fence typically costs $120–130. 9-gauge usually runs for about $180 for the same size roll. And that doesn't include the posts and rails. On average, 200 feet of chain link fence can cost somewhere around $3000 with installation.

Gates and Latches

Gates

If you are regularly needing to go in and out of a goat pen, ensure that the gates swing inwards only; because when you open it, you will move the goats away from the entrance with the gate. The goats are also less likely to push against the gate, and accidentally push it open to escape.

If you need to move goats into different areas often, it can be a good idea to have a gate for goats, and a gate for humans, because they will learn the gate they go into and any time it is opened they will try to get through it. It will also make bringing in wheelbarrows or tractors easier, because the goats won't be trying to make an exit purely out of habit.

Latches

Some people choose to put a number of latches on goat enclosures to keep them safe and to prevent them from getting out. Others choose to use a chain and padlock because this keeps the goats in and unwanted people out. Another option are chains with clips on metal gates.

How Much Fencing Is Necessary to Protect Your Goats?

If you have a goat or goats that regularly escape your fencing, then I would suggest that you erect taller fencing that they physically can't jump or climb over. You can stack panels and secure them with zip ties to make them secure.

If you have very friendly pet goats, then you may not wish to fence them in, but let them wander over your property almost like dogs or cats. But do beware if your goats come into heat, they can leap 10-foot-high fences to get to a buck.

It will also depend on the type of goats you have, and whether you have a mix of bucks and does that you want to keep separate until it's time for breeding. It will depend on how many goats you have, and the size of your property. It would be sensible to start with a few goats and gradually increase in size, rather than having an immediately huge goat herd from the start. You will need perimeter fencing and internal fencing for sectioning goats into different areas.

Keep in mind that if you decide to keep bucks on your property, you need to have a really solid plan for him once he turns into a breeding machine. Believe me, I've had my fair share of their hormone-induced antics. Not to mention the countless stories I've heard when I visited different farms. Most importantly, don't place your buck's area anywhere near does. If they share a fence, bucks will get to does somehow, or even impregnate them through the fence.

In conclusion, if you're just beginning to build your infrastructure, my advice would be to buy the best materials and fence you can afford one pasture at a time. Your goats will always keep trying your fence, and you will spend a lot of time, money and energy on fixing your fence and chasing your escaped goats. So, don't cheap out on your goat fencing even if it means keeping fewer goats.

When we first got our two goats, we used woven wire fencing, and I still think this is a really good option. It's very easy to put in place without any technical know-how. It's self-explanatory, effective, and you don't need a lot of tools. It's a fairly inexpensive, yet very dependable first choice option to keep your goats safe.

Key takeaways from this chapter:

1. Build your fences higher than you expect because goats are good jumpers. I typically recommend building a fence that's 5 feet high and not 4, as goats can often jump over 4-foot fences.
2. Make your fences really strong, because goats will lean against fences, clamber on them, and scratch themselves on the fence, weakening it.
3. Ensure the type of fence is suitable for the goats you have (i.e., that kids can't get through, or goats with horns can't get their heads stuck).
4. Make sure the gap at the bottom of your fence is not too big and that goats can't crawl under it.
5. Always have wire cutters at hand in case you need to get a stuck goat out of the fencing quickly.
6. Be prepared to make repairs and have spare fencing/boards at hand so you can do this quickly.
7. Consider gates that open inwards, and separate gates for humans and animals.
8. If you use electric fencing, then regularly cut the grass and weeds near it so that it doesn't drain the electric, and check for fallen tree limbs on the fencing.
9. It's worth training kids with very small electric shocks with the fence, whilst you are there to control the power and ensure that they do not get stuck in the fence.

So far the book has covered the different breeds of goats for milk, meat and fiber, goat terms, facts, and myths, where to get good healthy goats from and how to transport them home, the type of housing, bedding and troughs needed for goats, and the fencing to keep your goats where you want them and secure. The next chapter will look at the important topic of what to feed goats to keep them healthy.

Chapter 5: Feeding Your Goats

Goats are more like deer in terms of nutrition, rather than sheep or cattle. Goats browse rather than graze (cattle, sheep, and horses graze), but by browsing, goats will forage food from grass, plants, weeds, and shrubs. Goats can effectively clear overgrown land. Forage can be eaten by goats in the winter, and this can be either grass or alfalfa. Goats prefer to browse rather than to eat grass. Goats should never eat a diet of entirely fresh grass, as this would not give them the nutrients they require.

Generally, goats eat the following:

1. **Foraged foods:** branches, berries, herbs, and so on.
2. **Hay:** which has to be fed by a farmer, of course.
3. **Mineral supplements:** either store-bought or homemade concentrate.

Depending on how your goats have been raised before you purchase them may determine their feeding preferences. For example, a goat raised as a forager before you purchased it will be a healthy eater, like hay and anything green. But, if the people you bought your goats from raised them on grain, then your goats may now prefer grain.

Goats' Digestive System

We will look at goat health more thoroughly in Chapter 7 of this book, but one of the key health issues with goats is their digestive system. You do really need to look after them. Goats are ruminants, which means they eat plants, then these are digested in their stomach, which has four compartments to aid this. Hay acts as roughage for goats and they need this for their rumen (their first stomach compartment that contains live bacteria to start to the process of digestion) to work effectively. Because bacteria are fermented in the rumen, this creates heat. Goats have quite thin skin and don't have thick coats, so by giving goats forage this allows them to naturally produce their own heat inside to warm them up.

Always make changes to your goat's diet gradually. Because if you introduce new hays or feeds too quickly, this can cause digestive upset. The bacteria in their rumen, needs time to adjust to new food to help with effectively digesting it.

Goats do have a sensitive digestive system; you can give them baking soda along with their loose minerals and this can help their digestion and prevent bloat.

Goats' Nutritional Requirements

Foraged Food

Even though goats love grass, they still prefer some tree branches, a thicket of saplings, and herbs. Goats can eat the widest range of plants among livestock. They can digest things that other animals can't, and as a result, they can feed on almost any terrain.

Goats have always been a last resort animal for pastoral cultures. Even when land starts to turn to desert, they can still scrape together subsistence from whatever there's left. Goats are sturdy creatures, and when things get tough, they will do what it takes to survive.

However, you want your goats to thrive, not just barely survive. And one of the best ways to do it is to ensure they are being fed a healthy diet. I always suggest letting your goats browse for most of their food. This way, they can follow their natural instincts and save you some money at the same time. Letting them run wild in the woods or fields is not a good idea, of course. Ideally, you should have a large, fenced area, preferably with a guard animal, where they can browse.

If you don't have a large plot of wooded land, you can still let your goats enjoy feeding naturally. In this case, I would suggest walking your herd through woodlands at noon every day. It is quite a chore, and your goats can stray from you sometimes, which can be avoided with training.

But, if that is a bit too bothersome for you, you can still give them the benefits of foraging. You can bring them fresh-cut branches from trees and they will definitely enjoy them. If you do that, make sure to avoid wilted cherry branches, as they have a high cyanide content and can be poisonous to goats. Our goats always got excited when we were pruning out our fruit trees. If you bring some branches for your goats, make sure to hang them on the fence or from a gate instead of throwing them on the ground. Goats are naturally used to stretching up to eat their food rather than bending down to pick it up. Your goats will also eat fresh-cut herbs and weeds from your garden or around your land, if it hasn't been sprayed with herbicide, of course!

Hay

In case you don't have the land for your goats to forage or you don't have access or time to collect fodder for your goats, you can certainly feed them with hay. Legume hays, such as alfalfa, vetch, soybean, clover, or lespedeza are great for kids, pregnant does, as well as lactating does. Grass hays, such as Sudan, red top, timothy, bromegrass or fescue are less nutritious. A good all-purpose hay is a 50/50 grass-legume mix. Most importantly, never feed your goats moldy hay, it can make them really sick.

Goats need 2–4 lb of hay per day (3–4% of their body weight in pounds), they can have this to freely access, or they could be given this twice a day. If they don't have a lot to forage, for example, in the winter time, or if there's not much range access, then they need additional hay. Ideally, goats need 8 hours of grazing time per day. Alfalfa hay contains more protein, vitamins, and minerals than grass hay, so this is a good option for milk goats because it will give them more energy, and the calcium they need. You can also get lespedeza and clover hays also high in enriched protein and other nutrients.

Goats can be fed Chaffhaye which is young alfalfa or grass, mixed with molasses and a probiotic culture called bacillus subtilis, the hay ferments and this adds good bacteria to the goat's rumens. It's good to feed 2 pounds per 100 pounds of body weight for this.

If a goat's natural pasture area has things like millet, Bahia grasses, Sudan grasses, clover, sorghum, and grain grass mixture, these are all good healthy things that will ensure your goat is in the best health.

Goats perform at their best when they have been given diets that they can easily digest, that they gain energy from, and contain things like nitrogen and trace minerals. A diet like this helps the animal to grow, and to produce good milk if it's a dairy goat.

If the goats are given feed that has protein pellets in it, some goats will ignore the protein pellets and it's hard to know who has had what. But if the feed is all the same consistency (all pellets) they are less likely to ignore just the protein pellets and get everything they require from their diet.

Wethers (castrated male goats) should not be fed grain, bucks should not either, this is because having grain puts them at high risk of getting urinary calculi.

If breeders have fed does grain before you have them, then you may want to continue with this, and feed them ¼ to ½ a cup twice a day. Does for milking tend to eat grain whilst they are being milked.

Grain feed or a pelleted grain mix is beneficial to add extra protein, vitamins, and minerals into goat diets. Many farmers add to their goat's diet with some grain feed too, especially if a doe is feeding many kids, or if the weather is bad, and browsing isn't so much of an option. Grain shouldn't be overdone though, because you don't want to make your goats have an unhealthy weight or for them to develop illnesses. Feeding a goat too much grain can kill them. It's suggested they don't have more than 1.5 lb of grain per day and clearly kids should have less than this. It is suggested that you can add 12–16% grain to goat feed. Grain can include rye, oats, corn, and barley. You can get rolled oats to feed goats.

Nutrition for pregnant does needs to be increased six weeks before she gives birth to kids, this will ensure that when she gives birth, she has the nutrition levels she needs for lactation. When a goat starts to lactate, their protein requirements double. To create milk they need protein, so feeding a grain alone is not sufficient, but Alfalfa hay is the only hay with enough protein that will meet a lactating doe's requirements. The protein intake of a pregnant doe should be slowly increased as the pregnancy develops.

Making big sudden changes to a goat's feed can cause health issues. It is possible to purchase Alfalfa pellets too, but again, if you are switching your goats from Alfalfa hay to Alfalfa pellets, then do this gradually. Initially, they may eat a lot because they may assume they are treats. To provide more protein to a goat, this can include cotton meal, soybean meal, and fish meal. Buying hay in large quantities rather than just a few bales is less expensive, but if you do this, you will need a hay storage area or loft to keep it dry, clean, and safe.

What Else You Can Feed Your Goats

You can give your goats kitchen scraps that you'd usually compost, except eggshells and fish (though if you have chickens or ducks, they will eat them). Goats can be fed fish meal, as mentioned previously, but keep in mind fish meal is essentially dried, ground tissue of under composed whole fish or fish cuttings.

Goats will eat banana peel, orange peel, tomatoes, garlic skins, plus other fruit and vegetable cuttings. As a treat, you can give goats raisins or corn chips, but only a few, and goats like a slice of bread as a treat too. But be careful not to overfeed them. Other treats include sweet feed, and some human food, such as fruit, dried fruits, vegetables, Graham crackers, Cheerios, Cheetos, and corn chips. Similar to humans, too many snacks are not good for them. Goats will also eat weeds, they like plantain and kudzu.

Beat pulp is great for goats because it is high in fiber, protein and will give your goats energy. Another good thing to give excellent nutrition to goats are black oil sunflower seeds, these are high in vitamin E, and things like zinc, iron, and selenium. They will make your goats have healthy shiny coats, and make their milk have a higher fat content. Kelp meal is another thing great for nutrition, as it's a good source of iodine, and will help with milk production. You can add a little apple cider vinegar to your goats' water, and this may help boost their immune system due to the enzymes and minerals it contains.

Water

Goats need fresh water daily or twice a day. If there's any water remaining when you come to change it, then use it to water plants, never just top it up, it needs to be fresh. If there's any poop in the water, change it immediately. If you have a few goats, a 2-gallon bucket is sufficient. If you have 4 or 5 goats, then ensure you have two buckets. It's better to have several small buckets of fresh water, rather than just one large bucket. In case any of the buckets get poop or straw in them, then your goats will have other fresh buckets to choose from. A good tip can be to put the bucket higher, mounted on a fence so that your goats can reach to drink out of; but higher than your tallest goat's tail so that it can't poop in it. You could use a rain barrel to save water for your goats.

Supplements

If you feed your goats mainly or exclusively with hay, they need additional sources of nutrients which can be provided with concentrate, also sometimes called grain ration. It is a concentrated source of nutrients, proteins and minerals. You can get pre-made concentrate in stores or you can mix it yourself at home.

Pre-made concentrates usually contain soy, corn, bran, salt, minerals, oil meal, and molasses. Buying pre-made concentrate is perhaps the most convenient option, but it's neither the cheapest, nor the healthiest. Many homesteaders I've talked to over the years avoid pre-made concentrate because they are concerned about the presence of GMO corn or soybean oil meal, as well as pesticides. Despite that, store-bought concentrate is the most convenient option for many people.

You can make your own concentrate at home. You can use a lot of different ingredients, such as rye, barley, oats, sunflower seeds, pumpkin seeds, field, peas, mangel beets, wheat, and more. Making your own concentrate allows you to tailor your goat feed for specific needs.

You can pay to have a livestock nutritionist to formulate a concentrate that you can feed your goats when they need supplements. If you do this, it's advisable to pick one who is local to you, and they can specially consider the environmental conditions that your goats live in and assess what nutrients they will naturally get from browsing, and what nutrients they may be missing and need to be supplemented. The nutritionist can also test your hay, which will give more information on the nutrition the goats are getting and what needs to be added to the formula.

Alternatively, you can find a goat farm in your area. Most farmers or homesteaders would be happy to answer all your questions, including making your own concentrate for your goats. It's always best to learn from goat-keepers in your area, as they often will know how to best feed your goats in your environmental conditions.

Minerals

It can be a good idea to provide loose goat minerals which the goats can help themselves to freely, and a mineral feeder to put them in and for the goats to eat out of, such as Selenium-E and Sweetlix Meat Maker. Blocks are not so good, because these are combined, and goats don't really have the right sort of tongue to make best use of blocks. Phosphorous, salt and calcium are other minerals your goats may need, plus they need vitamins A, D and E.

Your goats' mineral requirements also depend on the area you are in. For instance, eastern and central United States are generally deficient in selenium, while some areas west of Great Lakes are low in iodine.

I would suggest talking to goat keepers in your area and learn how they handle their goats. Ask them how they feed their goats, what minerals and supplements they give them, and what diseases bothered them in the past. This can help you pick an optimal option for your own goats.

Poisonous Plants

Ensure you are aware of what plants are poisonous to goats, and make sure that they can't browse these. If goats eat toxic plants, they can get goat bloat, which can sometimes be deadly. Whilst goat bloat can be deadly, if you recognize the symptoms, you can give goats baking soda, which is very inexpensive to purchase. Try not to let your goats browse on a pasture that is rich in clover and alfalfa and is wet from rain or dew because this can cause bloat.

Plants which are poisonous to goats include:

- Nightshade
- Plum leaves
- Peach leaves
- Crotalaria
- Poke weed

Do check that you know what each of these looks like, and that the area your goats are browsing in does not contain them.

The most dangerous plant is Rhododendron—even a few leaves can be deadly.

Rhododendron

What Not to Feed Your Goats

If you are using your goats for milk purposes, I wouldn't give them too much garlic, or access to wild garlic, as this can impact the flavor of the milk. Onions and mint should be avoided for the same reason. While they are not harmful to goats, they will certainly affect the flavor of your goats' milk.

There are other things that you should never feed a goat, and these include: avocado, azaleas, chocolate, kale, holly trees or bushes, lilacs, lily of the valley, milkweed, cherries, rhododendron, and rhubarb leaves.

One important consideration is making sure there is no rat poison or herbicides anywhere near where your goats live and browse. The same is true for stomach-binding materials like old carpeting or tarps. Always keep these sorts of materials well out of reach of your goats.

We try to feed our goats as naturally as possible on our homestead. Our goats don't eat hay, concentrate, or minerals. However, not everyone has access to acres of wooden land. You can still raise healthy goats and feed them with hay, supplements and minerals. When it comes to feeding your goats, the most important part is taking time to do your research. I hope this chapter has provided you with useful information on how to feed your goats.

It takes time to get it right, so don't hesitate to ask fellow goat-keepers for advice and always pay attention to your animals. You'll have to experiment to get things right, you can make mistakes along the way, and that's fine. Learn from them and always strive to improve your methods of caring for your goats. The longer you live with these lovely animals, the better you'll get at understanding their needs and taking care of them. And the better you take care of your herd, the more enjoyable your life with goats will be.

Key takeaways from this chapter:

1. Don't make large changes to your goat's diet immediately. Gradually introduce new feed or hay so that it doesn't cause digestive upset, and their rumen (their first stomach) can start the process of digestion.
2. Goats require 3–4% of their body weight of hay per day, which is usually 2–4 lb of hay for a typical adult goat.
3. Alfalfa hay is good for goats and contains a lot of protein, which is great for milk goats and pregnant does.
4. Goats require minerals and vitamins for optimum health.
5. Goats can be fed chaffhaye which gives them useful probiotics.
6. Male goats should not eat grain, or they can get urinary calculi.
7. If you give your goat grain, it shouldn't be more than ½ a cup twice a day.
8. You can give your goat most fruit and vegetable scraps and some treats, but don't overdo these.
9. You can supplement your goat's diet with beet pulp, black oil sunflower seeds, kelp meal, cider vinegar, and other minerals.

10. You can have a goat nutritionist formulate a goat pellet for you that contains what they are lacking from their natural environment.

The next chapter will move onto goat care and maintenance, looking at grooming goats, taking care of their hooves, dehorning, disbudding, and tattooing them.

Chapter 6: Goat Care and Grooming

Goats are not too difficult to take care of, provided that you keep their housing clean and free from poop and pee, and that they have access to good nutrition and clean fresh water, access to pasture area, sunlight, and a cool breeze, then this should do a lot to keep your goats in good health.

Goats are skittish creatures and can be nervous, so it's best to be calm around them, don't shout, and be gentle with them. You can buy goat rope halters to lead goats to where you want them to be, and if you need to handle the goat for any other reason. Having another person to help can be useful, so the goat doesn't become distressed and starts to struggle. The more you handle goats whilst they're kids, the more they'll be used to being handled with ease.

Routine health care costs will include pesticides to prevent mites or lice on your goats (approx. $20). You can get Pyrethrin powder to rub on goats on their back, or Ivermectin Pour-On for cattle. You can buy ophthalmic ointment to treat pink eye ($20) and dewormers (again about $20).

Goat Grooming Supplies

In order to adequately groom a goat, you need a curry comb, hard and soft brushes, a comb for beards and tails, and hoof trimmers. If your goat is mucky, then a bath mitt can be useful too. Electric clipper can neaten up their tails or general appearance.

Brushing

Grooming your goat allows you to check on the health of your goat (see Chapter 7 for more information about goat health). You can get rid of any mud on the goat, by brushing it with a hard brush, if there are less obvious gritty bits in the goat's coat, then the curry comb will bring these out and it will give the goat a massage too. The soft brush should be used to finish off the grooming and it will distribute natural oils through your goats' coats.

Grooming your goat allows you to check on the health of your goat (see Chapter 7 for more information about goat health). Brushing your goats is a good time to also feel their bodies with your hands and look for any bumps and lumps. These could be a parasitical skin infestation or wounds. Be careful and don't brush your goats too hard in wounded places.

You shouldn't need to bathe a goat, but if your goat has parasites, this may be an exception to get rid of lice and help with clipping a goat. The water should be slightly warm when bathing a goat. Washing a goat is not difficult at all, just wet the goat, use an animal or goat shampoo, and rinse.

Hoof Care

Goats' hooves should be trimmed every 4–6 weeks, and I can't stress enough to set up a firm schedule for when this will be done to maintain them, it can be done on a regular calendar, diary, or an app on your phone or computer. If hooves aren't looked after, goats can become lame or their hooves can contract an infection. If you start the practice when they are young, they will be less resistant to it, but they are still not keen on having their back hooves done. You can get a person to stand at their front to distract them with a treat whilst this is being done. You can also use a milking stand to keep the goat in place whilst it's having its hooves trimmed.

If you feel confident enough, and have a number of goats, it would be beneficial to learn how to trim your own goats' hooves, with some $30 trimmers, rather than repeated expensive trips to the vets for this. A vet should be able to show you how to do this. If possible, it's always best to learn from a veterinarian or an experienced goat-keeper. There are nuances in trimming technique that cannot be conveyed through words alone.

Before you get to trimming your goats' hooves, it can be useful to have some corn starch handy in case you accidentally trim the hoof too close and cause it to bleed, you can apply corn starch and a dab of antibiotic ointment. It's sensible to wear thick gloves when hoof trimming, as the clippers are sharp, and the animals will resist and make movements whilst trying to do it. When you are taking care of hooves, ensure there are no rocks or sticks stuck in the goat's hoof, and also check that their hooves don't smell, as this can be a sign of hoof rot (See Chapter 7 for more information on this)

As for the trimming procedure, it's not too different in concept from trimming dog's nails. Just like cats or dogs, goats have a sensitive area made of soft tissue in the center of their hooves, which is called the quick.

Once your goat allows you to raise their foot relatively calmly, use a brush to clean any surface dirt off their hooves. Then, take a clean pair of hoof trimmers and use the tip to clean out the dirt from your goat's hoof, especially between the hoof wall and the soft sole of their feet. It's highly likely there will be a lot of dirt and debris to dig out.

I usually recommend that you start trimming at the very front of their toes, but it depends on how overgrown the hoof is. Start clipping the overgrown part of the toe tip just a bit at a time. As you keep trimming the hoof, you'll see that the surface of the remaining hoof wall starts to turn white (or black, if your goats have black hooves). Always trim only a little bit at a time. If you see any pink areas, stop immediately, as it usually means that you're approaching live tissue that will bleed and cause your goats pain if you cut it. Sometimes, even if there was no blood, you might have still trimmed a bit too far, which can cause some discomfort when walking on certain surfaces.

But what if the hoof is overgrown and folded over the sole? In this case, you should start trimming at the outer side of the hoof, trimming the flap away. Work around the hoof wall until you reach the back of your goat's foot. Sometimes, you may need to trim the heel, but be careful, as this is a very sensitive area.

When you've leveled the hoof, take a look at the dew claws. These are the claws protruding from above the hoof on the back of your goat's leg. They usually require minimal trimming if they are maintained regularly. However, the older a goat gets, the more often you'll have to trim their dew claws.

Once you've trimmed all four feet, let your goat walk freely and watch them walk. Pay attention to their motion, they should be walking normally. This way you can determine if your trimming needs some adjustment for their comfort.

Goat Care and Grooming

Below you can find a step-by-step chart for goat hoof trimming procedure that will hopefully help you better understand the basics of goat's hoof trimming.

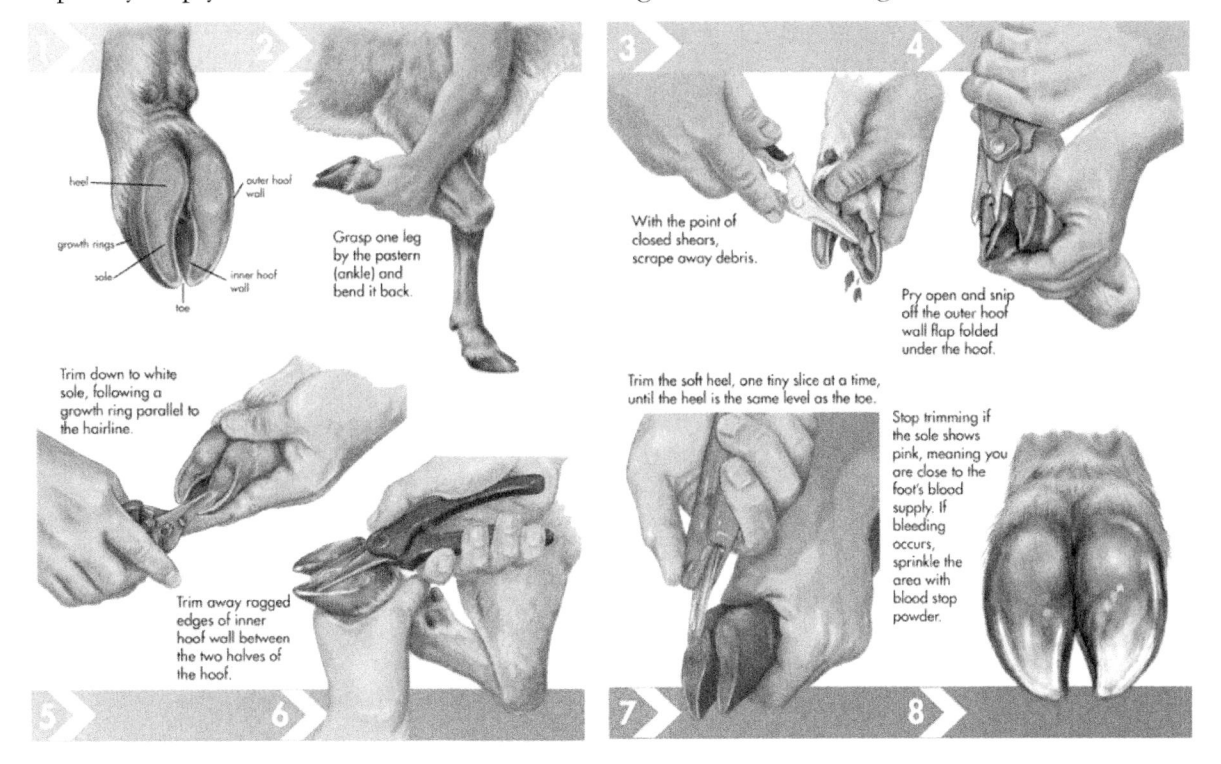

1. heel, outer hoof wall, growth rings, sole, inner hoof wall, toe

2. Grasp one leg by the pastern (ankle) and bend it back.

3. With the point of closed shears, scrape away debris.

4. Pry open and snip off the outer hoof wall flap folded under the hoof.

5. Trim down to white sole, following a growth ring parallel to the hairline.

6. Trim away ragged edges of inner hoof wall between the two halves of the hoof.

7. Trim the soft heel, one tiny slice at a time, until the heel is the same level as the toe.

8. Stop trimming if the sole shows pink, meaning you are close to the foot's blood supply. If bleeding occurs, sprinkle the area with blood stop powder.

Tattooing

All dairy goats must have a tattoo before they can be registered with the American Dairy Goat Association. Every goat, except LaMancha goats, should be tattooed in the ears. The LaMancha goats should be tattooed in the tail web area. The tattoo should be up to 4 letters or numbers, but you can't have one letter followed by a number or numbers. You can't have K2 or D347, for example. You need to tattoo according to your membership identification number, and you can't use a tattoo that someone else has been given. The tattoo should be in the right ear, right tail, or center tail. You should tattoo your goats before they are sold or purchased. In the goats' left ears, you can use a letter that represents the year of birth, i.e., L = 2019; M = 2020; N

= 2021 and then, depending on what kid is born into the herd, that can be given a number, i.e. N1, N2, N3, etc.

When tattooing a goat, it should have a muzzle or halter on it. The area to be tattooed should be cleaned with alcohol. You need to make sure the right symbols are set up, then press the thin rubber sponge pad down firmly over the needles. You can check the symbols are correct using paper, then you smear ink on the skin and make sure you are piercing in between veins and cartilage. Green ink is best to use for permanence.[6] Make the imprint quickly and firmly, rub more in into it. You could use an old toothbrush to work the ink in, then let the area heal, which usually takes around 1–3 weeks. Ensure that you keep track of the tattoo numbers in a record. If you are struggling to read the tattoo of a dark eared animal, you can use a flashlight on the outside of the ear.

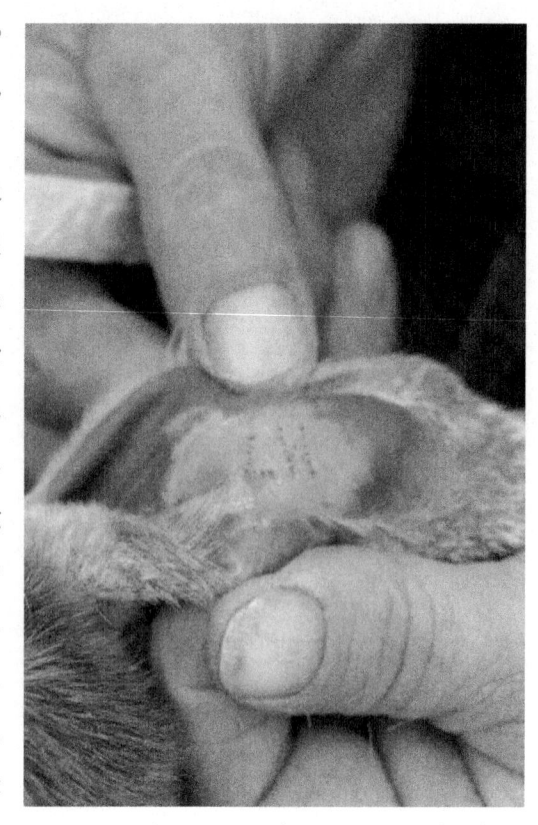

Shearing

If you have goats specifically for their fiber, such as Angora or Pygora goats, then they will need shearing twice a year. If the goats are only sheared annually, it can mean that the fiber gets matted in the spring. You can learn to shear the goats yourself, but it is really time consuming and can make your back ache, or you could hire a professional shearer and they will do this much quicker. Experienced shearers can do around 14 goats in one afternoon. If your goat has long hair, by having them clipped or sheared, it can make them feel more comfortable. Longer hair is fine in the winter to keep them warm, but not so comfortable in the warmer weather. You may

---

[6] Image from alifeofheritage.com: https://alifeofheritage.com/farm-living/how-to-tattoo-a-goat/

want to ensure that in the winter their coat isn't so long that it gets mud clods attached. Chapter 11 will go more in depth on the procedure of shearing goats, if you'd like to start keeping goats specifically for fiber.

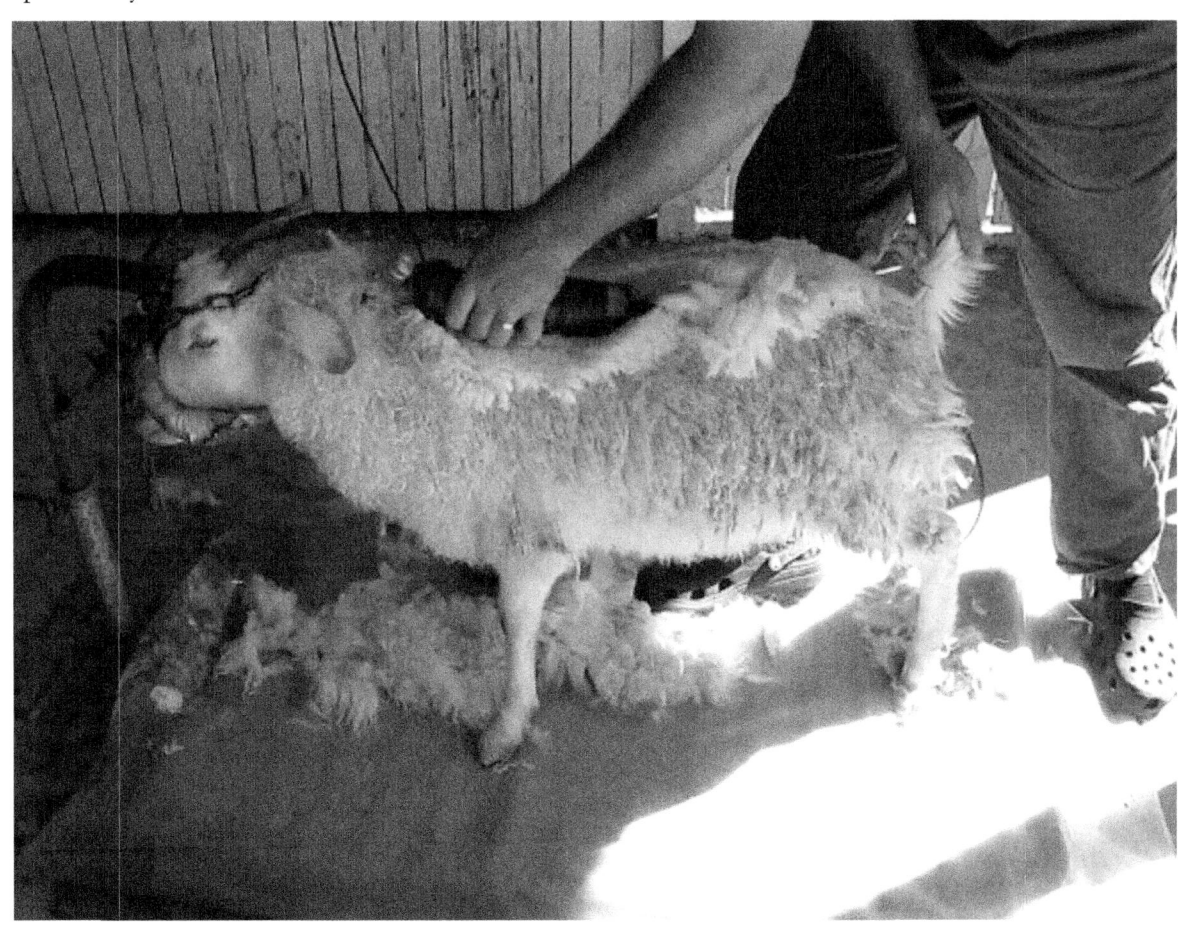

Leash Training

It is best to start leash training your goats when they are young. If you're constantly around them, be it handling goats, feeding them or working around them, they will naturally see you as the leader of the herd.

Put a collar and a lead or a halter on the goat you want to train. You can also use a dog harness, it is the easiest and safest solution, as it won't choke your goats like a collar can. Start out by introducing them to the idea of halter or harness. Put it on them when you're around, let them sniff it and play with it. Don't leave it on when goats are not under your supervision, however, as it can catch on to things and goats can hurt themselves.

When they get used to wearing a halter or harness, start in a smaller area and walk forward a few steps. If the goat follows, continue walking. Stop every few steps and reward your goat with a treat. Gradually increase the distance between stops. Lead the goat and do not let it lead you. If they try to stray away, say "stop" or "get back".

Do not drag the goat and tug on the collar. It can block the windpipe and cause the goat to collapse. They may drop to their knees, but they will recover quickly. Never drag your goats, but gently lead them.

Once they feel more comfortable wearing a halter or a vest, you can take them out into a larger area. Remember to take the halter or vest off when you finish training your goats.

We have goats for milk, fiber, and meat. We have everything down to a fine art now regarding the goats' schedule to ensure they don't have lice and have been dewormed. We groom them with brushes and combs each week, and while we're doing that, we can check the goats' skin and ensure all is well. We trim their hooves every 4 weeks and check them for stones and hoof rot. We have our goats tattooed, and we shear our Angora and Cashmere goats twice yearly, and we're able to make beautiful products out of the soft wool from them.

Key takeaways from this chapter:
1. Be calm and gentle around goats.
2. You will need to use products to prevent lice, and to deworm your goats.
3. There are different combs and brushes required to groom a goat.
4. It can be a good idea to put together a goat first aid kit.
5. You need to trim goats' hooves every 4–6 weeks and check for hoof rot.
6. If you don't want your goats to have horns, they should be disbudded before they are 8 days old, apart from Nubian does around 14 days. Goats should be given anesthetic and pain relief to make this as pain free a procedure as possible.

7. If you want to register a dairy goat, it will need a tattoo that matches your membership number, and most goats are tattooed in their right ear, except LaMancha goats, who have their tail area tattooed.
8. If you have goats for fiber, they should be sheared twice a year.

The next chapter of this book, will move on to the very important topic of goat health, looking at how to keep your goats healthy, how to find a vet, common issues that could crop up with goats that you need to look out for; and how to tell when your goat is sick.

Chapter 7: Goat Health

Basic Healthcare Requirements

Any time throughout the natural course of the day, when you are feeding your goats, changing their water, topping up minerals, or cleaning their shelter, always take special care to check on the health of your goats. You can ensure that they still have a good appetite, make sure they do not have a limp, ensure there are no breathing difficulties, make sure they don't appear to have diarrhea, look at their eyes and nose and ensure neither have any discharge from them, you can feel their temperature and even more accurately check it with a thermometer to ensure it is normal. All of these things will give you an insight into the health of your goats.

I would strongly recommend that you keep a record of your goats, their breed, age, weight, temperature, any health supplements they take, and so on so that if you ever need any veterinary assistance for them, or anyone else to care for them, they know what is "normal" for your goats and know what they've been eating and about any medications your goats usually require. I have created a Goat Record Keeping Log Book that will allow you to keep all the important information about your herd in one convenient place. If you're interested in getting a log book, you can find out more by going back to page 5 in the beginning of the book right before the table of contents.

If you have pre-existing goats and you are bringing more goats onto your farm, you need to quarantine the new goats for four weeks. Don't wear the same clothing around your new goats and your existing ones. Always change your clothes after seeing your new goats so that you do not spread any diseases amongst your whole herd. Always wear disposable gloves when checking over new goats.

Depending on where you are situated, vets don't always have a lot of experience with goats, so you may end up doing a lot of health care for them yourself.

Vaccinations

There are various vaccinations that are beneficial for goats to have. These include:
- Rabies

- Tetanus
- Clostridium (CDT): Pregnant or nursing mothers may need a booster shot of this if they've been isolated for more than 4 weeks. Kids should be 1-month-old before being vaccinated for this.

The rabies vaccination can make the goats feel rough, so it is best for the goats to have had the other vaccinations at least a month before, or after their rabies, and not all at once.

Deworming Goats

Dewormers are chemicals or drugs used to remove worm parasites. They are also called "Anthelmintics". Dewormers have been used for effectiveness and safety for use in animals, and typically pharmaceutical companies will not market a dewormer unless it is essentially 100% effective. Only FDA-approved dewormers can be used without any restrictions. All other dewormers are extra-label and are subject to specific regulations as outlined by the FDA. Goats can develop dewormer resistance, so using dewormers correctly is important to avoid this.

As to when you should deworm your goats, the short answer is when they need it. Goats are resistant to many deworming products, so you don't want to deworm them when it's not necessary. One instance when you definitely should deworm your goats is deworming does before kidding. Otherwise, there are a number of different techniques to check if your goats need deworming.

You can get samples of their fecal matter and check it for parasite eggs, but it takes a few weeks and can cost quite a bit. Also, fecal egg counts by themselves are not a very reliable way to determine the worm load an animal is carrying. Plus, there aren't any agreed upon thresholds for treatment. So, it is best used in combination with other criteria. Still, fecal testing is important. You can talk to your vet and organize to get fecal samples sent to a veterinary lab to get them tested. This will help determine what parasites your goats have (goats typically have tapeworms, lungworms or barber pole worms) and what dewormers you should use to treat them.

You can check your goat's mucous membranes by pulling down their lower eyelid. They should be a nice pink color. If they are nice and pink, you don't need to deworm your goats. But if they are pale pink or white, you need to deworm them. Other symptoms of parasite infestation include pale gums, lowered eyelids, rough coat, and runny stool. Generally, it's always best to consult with your veterinarian on whether your goats need deworming and what products you should use. Also, your vet can help determine which animals need deworming using FAMACHA score, which measures anemia by comparing the color of the lower eyelid to a color chart, and is especially valuable when dealing with the barber pole worm. The Five Point Check is another technique and utilizes the FAMACHA score, but also includes scoring body condition, hair coat, soiling around the tail, and edema under the jaw to detect worm loads from parasites besides the barber pole worm. Using the FAMACHA method requires certification and most vets would have that. However, you can get online FAMACHA certification yourself if you wish.

There are three categories of dewormers. Drugs in the same category work in a similar manner and are used to kill similar worms. The three categories of dewormers are benzimidazoles, imidazothiazoles and macrolides. Let's take a closer look at each of them:

1. **Benzimidazoles or white dewormers**

Benzimidazole dewormers are sometimes called white dewormers because they are typically sold as a white liquid or paste. The more commonly used Benzimidazole dewormers are Fenbendazole (Safe-Guard, Panacur) and Albendazole (Valbazen). It may seem that there are more dewormers than there actually is because different brand-name drugs have the same one or two active ingredients. Only Safe-Guard is approved for use in goats, but other dewormers are used off-label.

White dewormers are only effective against tapeworms. A single dose of Albendazole kills tapeworms, but Fenbendazole must be given three days in a row to treat tapeworms. Also, Albendazole is not recommended for use in early pregnancy, as it may lead to problems at different stages of gestation.

2. **Macrolide or clear dewormers**

The second category of dewormers is Macrolide dewormers, also known as clear dewormers because they typically come as a clear liquid or gel. Macrolide dewormers are

Ivermectin (Ivomec) and Moxidectin (Cydectin). Moxidectin is much stronger than Ivermectin, so Ivermectin is usually used first so that you can use Moxidectin later if some goats develop dewormer resistance and Ivermectin no longer works. These drugs are not labeled for use in goats; however, they are approved for extra-label use orally or as a pour-on. A pour-on must never be used orally.

3. **Imidazothiazole or solid dewormers**

The third category of dewormers is Imidazothiazole dewormers or solid dewormers because they typically come in a solid form: water-soluble powder, a bolus, a medicated feed, or a feed additive. Imidazothiazole dewormers are Levamisole (Levisol, Tramisol) and Morantel Tartrate (Rumatel). Only Morantel tartrate is approved for use in goats, and it is sold as a feed additive.

Levamisole is rarely used in goats, and because of that it can work when other dewormers are no longer effective, including Rumatel, which is in the same class. Levamisole comes as a bolus or as a powder that can be mixed with water and given orally. Be careful when using levamisole because the margin of safety when dosing is not as large as with other dewormers, so it's easy to overdose a goat. Most sheep dewormers can be used in goats at twice the sheep dosage. Levamisole, however, should be used at only 1.5 times the sheep dosage. The signs of overdose are excessive salivation, tearing of the eyes, and the goat walking like it's drunk. Most likely, the goat will recover after a few hours or days of rest, but if the overdose is too high, it can cause death.

Castrating Goats

You don't need many bucks to breed a herd. One buck can sire dozens of kids, so you might want to choose the ones who will breed and castrate the others, unless you're planning to butcher them within a couple of months. Otherwise, keeping a lot of bucks can be troublesome, as they during the breeding season they start to stink and pee on themselves, and fight among themselves a lot too. Ideally, you should castrate your goats when they are 8–12 weeks old.

There are three different methods of castrating goats, but banding with an elastrator band is the most common one. It's a fairly easy procedure, but if you don't feel confident, you can have a vet or an experienced friend castrate your goats. You will need someone to hold the goat, so you can hold it and watch the procedure. Here's what you'll need:

- 1 cc Tetanus shot
- 0.25 cc Banamine pain reliever
- Castrator rings (keep them in the fridge so that they keep their shape better)
- Elastrator
- A clean sharp knife or a scalpel

Here are the steps you'll need to follow to castrate your kids with an elastrator band:

1. 30 minutes before castrating a kid, give them Banamine and Tetanus shot
2. Gather your supplies and get the kid in the proper position. As mentioned previously, you'll need someone to hold the kid. This person can either hold the kid's head in their lap and point their hindquarters towards the person operating the elastrator. Or they can grab the kid in the midsection, lifting their hindquarters into the air and allowing the kid to support themselves on their front legs.
3. Once the kid is in position, hold the tool to where the ring is facing the goat kid. The ring is opened by squeezing the handles of the elastrator. Place the castration ring over the testicles and scrotum. Make sure you get the ring around both testicles and they're below it before removing the castration ring from the elastrator. Ensure it's placed snuggly against the body. Once you make sure both testicles are placed correctly, stop squeezing the elastrator and remove the prongs from the band. Do a final check to make sure both testicles are beneath the castration ring, and nothing shifted when removing the elastrator. If you notice that the band is not on correctly or if one of the testes is not below the ring, cut off the band and repeat the procedure.
4. The ring will cut off the blow flow to the testicles, and you will see them begin to dry up. The banded parts should fall off in about two weeks. Sometimes a loose piece of

skin will remain attached and keep the male parts dangling. If that's the case, use a sharp, clean knife or scalpel to slice the piece of skin and remove the parts. Don't do this prematurely, however. If three weeks have passed, and the testicles are still barely hanging onto the goat, go ahead and use the scalpel to remove them.

Disbudding and Dehorning

Before you decide whether you should dehorn your goats or not, you should understand the purpose of a goat's horns.

For goats in the wild, their horns serve a very important purpose—protection for the predators. A horned goat can easily headbutt a predator and knock them off or stun them, which will give them enough time to escape.

Horns are also a part of a herd's hierarchy. As mentioned previously, there are always dominant goats in a herd, both bucks and does (a King and a Queen). Goats determine who will be a King or a Queen by head-butting each other, and then a King or a Queen they can defend their position if they are challenged by another herd member.

Horns are not only used for attacking or defending, they also have blood vessels in them that help regulate the body temperature. A goat without horns can still be fine in the hotter weather, but they might be a bit more uncomfortable.

With that said, here are the pros and cons of keeping a goat's horns intact:

Pros

- Horns give you the ability to control your goats. They give you something to grab on to and they can be great handles, if you're careful.
- Horns aid in regulating your goats' body temperature and help keep your goats cool. If you live in a hotter climate, you may consider keeping your goats' horns intact.
- Goats used their horns for self-defense. It's usually not an issue for domestic goats, and you can keep a guard dog to protect your goats, but you can consider leaving your goats horned if there are predators in your area.
- Disbudding your goats can cause brain injury or even kill your goats.

Cons

- Goat horns are used to defense and to determine the leader of the herd. Consequently, goats can cause damage to fellow goats and people. A blow from a horned goat can be really painful. Goats normally don't headbutt people, but bucks in rut can get aggressive and try to challenge you.
- If you keep your goats horned, you'll need to make special fencing considerations, as mentioned previously. Any openings in your fence would have to be no more that 4x4 inches, otherwise goats can get their heads stuck in the fence.
- If you plan to show your goats, you will have to disbud them, as many goat shows require that for the safety of handlers and other goats. We're never been interested in showing our goats, so if you don't plan on showing yours, it shouldn't be a concern.

Disbudding is a term used for a goat that had its horns removed at a very young age. Sometimes small pieces of horn can grow back if they weren't fully removed during the disbudding procedure and if this is the case, these are known as scurs.

If you want to do early disbudding, you can get a veterinarian to do this, or once you've been thoroughly trained, you can do this yourself, after purchasing a dehorning iron for around $100.

It's a job that people on farms tend to dislike doing, but it can be important to dehorn goats to prevent them from hurting other goats in the herd, or handlers. In the wild, goats would use their horns to fend off predators, but when they are on a farm, they usually don't need this. If they have horns, they can cause injuries to other goats or get them stuck in gates and fences. Most dairy goats are usually disbudded (this means destroying the tissue that would grow into horns).

After a goat has reached 3 weeks old, the tissue that would develop into horns attaches itself to the skull of the goat and a nub of horn can be felt. At this point, it is much more difficult to remove the horns. This is now called dehorning rather than disbudding, it is much more invasive and has more risk to the goat's health.

Therefore, it is better to disbud and allow time for that, rather than dehorn.

If an adult goat has horns or scurs, these should not be dehorned. If the goat is causing issues with its horns to other goats or handlers, then try to segregate horned and hornless goats. If a horn on a goat ever does need to be removed, you will need a veterinarian to do this, anesthesia to be administered, and the goat afterwards will be at risk of infection. Banding horns is strongly discouraged, as this can take 8 weeks, and causes the goat pain for that time, which is cruel and unnecessary.

Generally, doe kids, for example Alpine or Saanen should be disbudded at 5–7 days old. Buck kids should be disbudded at 3–5 days. Nubian doe kids should be disbudded at 14 days.

Most vets tend to use hot iron disbudding, rather than disbudding caustic paste. The goat can be given analgesics to reduce pain, there is a risk of damaging the goat's brain, but if proper training has been given, it is a proven method. Caustic pate can rub off on other goats or other parts of the goat's body, which can cause unintended burns, and it can cause prolonged pain. Disbudding can be painful, so it's most assuredly best to give the goats an anesthetic to numb the area and pain relief once the anesthetic wears off.

Disbudding Procedure

Disbudding a goat is not the most pleasant job in the world. You will need a kid holding box, unless you disbud your kids under anesthesia. You can have someone help you by holding the kid's head with their ears folded back, but you can do it all alone. You will also need the following supplies:

- Disbudding iron

- A knife to cut off the buds after ironing them.
- Electric clippers or scissors to trim the hair around the buds.
- Tetanus shot: a syringe with 1 cc of tetanus antitoxin.
- Banamine, it's a pain reliever and anti-inflammatory which you get from a vet. Aspirin or ibuprofen will also do the trick, but you need to give them with food.
- An antiseptic spray, you can get Blu-Kote at a feed store, for example.
- If the kid is bottle-fed, a bottle to comfort it following the procedure.

And now, for the disbudding procedure:

1. Give the kid the tetanus shot and pain reliever. The pain reliever usually takes about 30 minutes to work.
2. Plug in the disbudding iron at preheat until it becomes red hot. It usually takes 10–20 minutes to preheat it.
3. Meanwhile, trim the hair around the buds using electric clippers or scissors.
4. Place the kid in the kid holding box. If you're doing it by yourself, make sure that the ear nearest the horn bud is folded back and tucked into the box. If you have a helper, they should hold the kid's head firmly with the ears folded back.
5. Before you begin, test the iron on a piece of wood. If it burns a nice ring within 2 seconds—it's ready.
6. Place the disbudding iron to the horn bud evenly. The open circle goes around the tip of the bud. Don't press it too hard - the weight of the iron should be your pressure. Slowly rotate the iron around in the clock-wise direction for 3–4 seconds if it's a doeling or 5 seconds if it's a buckling. You want a nice copper ring around the base of the bud.
7. Now cut off the bud with a knife. Make sure to get low enough to the base of the bud. It should come off really easy.

8. The base of the bud may bleed. If it does, use the side of the disbudding iron and burn the base of the bud lightly to cauterize the wound. Move the iron around to a good burn on the top.
9. Now do the same thing for the other bud.
10. Before you're done, you'll need to go around once more on the base of each bud for 3 seconds.
11. Check the buds carefully and make sure you didn't miss any spots. If it all looks good, spray some antiseptic on the buds while making sure to avoid the eyes. You can bottle-feed the kid to comfort them or put them under their mother to nurse.
12. Check the buds the next day. There may be a little blood, and that's okay. If they start bleeding, you should cauterize them again, but you most likely won't have to. After a couple of weeks, the hair will grow back and you won't see any horns.

How to Find and Use a Veterinarian

Try to find a vet who is local to you, who can look after goats. Sometimes vets don't work with farm animals, they're more used to dogs, cats, hamsters, and so on. You may need to look for a larger animal hospital. The AG Service Finder website allows you to put in your city and state to find a goat vet near to you. Get the vet to visit the farm at least once a year to check the entire herd's wellness. The vet will then have a better idea of how you manage your herd and what their needs are. If you have any minor issues, you should be able to call your vet about them, but they can come out if it is an emergency. Goat veterinarians can be difficult to find in some areas, so this is something well worth doing your research and looking into prior to getting goats. The American Association of Small Ruminant Practitioners (AASRP) can help you find a goat veterinarian close to where you live. If there are other people with goats or sheep near you, you could ask them for recommendations too.

Common Goat Health Issues

Below are some of the common diseases that goats can contract. It is always sensible to test animals for disease before you bring them home and then test them again once you have them home. This can ensure that you are not bringing unhealthy goats near to your healthy ones and spreading disease.

Abscesses

Goats like most animals can get an abscess, which are typically caused by a foreign object, such as a splinter or a thorn, lodging under their skin and becoming infected. A lump may appear on the goat's body. If you notice one, ensure to put your goat into isolation away from other goats, and contact a vet. Because if a goat has an abscess, it could just be that, but it could be a sign of Caseous Lymphadenitis (CL). You don't want any diseases spreading amongst your goats.

Brucellosis

It is caused by a bacterium called *Brucella melitensis*, and this can cause pregnant goats to have an abortion at approximately the 4th month of their pregnancy. It can also cause arthritis and orchitis (inflammation of one or both testicles). If a number of your goats start having abortions, or swollen udders or testicles, and look nervous, and have a fever, these are typical symptoms of Brucellosis. Humans can catch it too, and get a fever, headaches, muscle, and joint pain and be extremely tired. Humans can contract it from handling animals who have brucellosis or consuming raw milk/meat from infected animals.

The good news is, B. melitensis is rare in the United States. However, there is no effective treatment for Brucellosis, unfortunately.

Caprine Arthritis and Encephalitis (CAE)

It's a contagious viral disease and goats catch this usually from their mother, and the colostrum in their milk. Once animals have this, they are infected for life. However, even if they are CAE positive, they can live with the condition for years and years and live a happy life. We have a goat who has this, who is over 12 years old. It can be passed via blood and feces, but we've never had any issues with other goats catching it.

There are 5 major forms of CAE in goats: arthritis, encephalitis (inflammation of the brain), pneumonia, mastitis, and chronic wasting. The arthritic form of the disease is most common in adult goats, while the encephalitic form is most common in kids. The chronic wasting form of the disease can occur either separately or in addition to any other form of CAE. Arthritis can come on suddenly, but in most of our older goats can show signs of arthritis from the age of 7 onwards, some joints can be a bit sore, swell and fill with fluid. Goats with arthritis may stand less, and lay down more, they could have a limp, they may lose weight and their coat can look dull.

CAE can be tested for with an Elisa blood test. There is no specific treatment for CAE. However, goats may be given supportive care including pain medication and antibiotics for opportunistic bacterial infections. Even with supportive care, the encephalitic form is usually fatal.

CAE infection and spread may be prevented by purchasing only test-negative animals or maintaining a closed herd and removing kids from infected does immediately after birth. Kids should receive only heat-treated colostrum and pasteurized milk or milk replacer. Individuals testing positive for the CAE virus should be removed from the herd.

Caseous Lymphadenitis (CL)

It's a chronic disease which causes large but non-painful abscesses on the skin, organs, and lymph nodes. Goats can get internal and external abscesses. If they have the external version, this can become internal via the blood or lymphatic system. If your goat has large abscesses, do get it checked by a vet for CL—this is the best way to test. Separate the goat from the herd so that it doesn't spread the disease, as it is contagious. You should check your goats each week to see if they have any lumps on their skin, and if it's good weather, you could shear the goats to better see their skin condition.

There is no cure for CL. However, CL abscesses must be treated to prevent ruptures and further contamination of other animals and environments.

Chlamydia

Chlamydial infection can cause abortions in goats. Chlamydiosis or Chlamydophila abortus, is a disease without symptoms and unfortunately, it's virtually undetectable in a herd until multiple abortions occur. There is no general pre-breeding screening tool for does; however, it can be tested for in bucks' semen. It can be spread via reproductive fluids, aborted tissue of infected animals, and carrier animals born to infected animals. Pasture and bedding can also be contaminated and remain so anywhere from a few weeks to a few months, depending on environmental conditions.

Diagnosing Chlamydia in goats is done by laboratory testing of placental tissue. Blood tests are not reliable unless they are taken at the time of abortion and again at three weeks. Chlamydia can be treated with Tetracycline or Tylosin, or other effective antibiotics. You should consult your vet for an appropriate course of treatment. There is no chlamydial vaccine for goats; however, the vaccine for sheep is relatively effective. Control measures include practicing good sanitation and establishing an effective vaccination program. Buy replacement does and kids from reputable sources with no history of the condition. Be aware that Chlamydia is contagious to humans.

Coccidiosis (Cocci)

This is a sickness that can cause diarrhea. Coccidia are protozoan parasites, most adult goats won't be bothered by these, but kids can become very sick and even die from this. Coccidiosis usually affects goats between 3-weeks-old and 5-months-old. It's a good idea to have a fecal sample from your goats analyzed quarterly in a year to check it for parasites. If you regularly rotate goats from pasture to pasture, it will prevent the buildup of worms. If you see kid goats with diarrhea or pale feces and the kid isn't growing well, it could be a sign of this, so get the kid checked with the vet.

Coccidiosis can be treated by drenching goats orally with amprolium (Corid 9.6%) for 5 consecutive days. This is often considered an effective form of treatment. This is an extra-label use, as amprolium is not labeled for goats, and a veterinarian needs to prescribe its use.

Sulfa drugs (sulfadimethoxine-sulfamethazine), such as Albon and Sulmet, are most effective in the early stages of acute infections when coccidia are multiplying rapidly. Sulfa drugs may not cure coccidiosis but are often given to infected goats to prevent secondary infections, such as bacterial enteritis. This is also an extra-label use, as sulfadimethoxine and sulfamethazine are not US FDA labeled for goats, and a veterinarian needs to prescribe its use.

Diarrhea

The most common cause of diarrhea in goat kids between 3-weeks-old and 5-months-old is coccidiosis. Diarrhea in adult goats is not common. One of the most common causes of Diarrhea in adult goats is worms. It can also be caused by a sudden change of diet and stress.

Keep in mind, diarrhea is a symptom, rather than a disease. We rarely give our goats any type of anti-diarrhea medication because they don't do anything to address the cause of the diarrhea. Just because it's stopped doesn't mean the goat is cured. If a goat seems healthy and happy and is in good condition otherwise, I would suggest you wait to see if the diarrhea stops by the next morning. If not, you should at other symptoms and contact your vet. They will be able to provide a proper treatment for cocci or worms.

Enteroxemia

Also known as overeating disease. It usually affects young goats on concentrate feeds. Goats are typically affected with a hemorrhagic form of enterotoxemia. Signs include sudden loss of appetite, bloat, lack of rumen activity and rumination, depression, your goats can walk as if they're drunk. As the disease progresses, the goat becomes unable to stand and lies on its side. Fever and diarrhea can also be signs of Entero.

Anti-toxin, anti-inflammatories, oral activated charcoal and probiotics can be helpful very early in the course of the disease. At the first sign of an enterotoxemia outbreak in a herd, the remaining kids should be given injections of C and D antitoxin and a C and D vaccine booster.

Goat Bloat or Grain Poisoning

Goats usually pass gas as they digest their food, releasing the fermentation gases. Bloat happens when the gases form in tiny bubbles and the goat is unable to pass them.

There are a few things you can do to prevent goats from bloating. Don't place goats into a new pasture and allow them to overeat there. Their digestive system needs time to adjust to any new food. If you have fields that grow alfalfa and clover, these are good, but let your goats have a bit of alfalfa and clover with their food slowly increasing this over time before just placing them in a field full of it. As mentioned previously, don't let them into a field that is wet from rain or dew, as this has a greater likelihood of causing bloat. Do ensure that any grain storage containers are totally secure because if a goat can get into them, they will gorge themselves on it, which can lead to bloat and poisoning.

Goats can get either free gas bloat or frothy bloat. Free gas bloat is usually caused by an obstruction in the throat or esophagus that prevents gas from escaping. If you can see or feel a blockage at the back of your goat's throat, you may be able to remove it with care. Also, if you see a bulge on the left of the neck, you can try to massage it down gently. If you already have experience, you can pass a stomach tube down the esophagus. This will quickly relieve free gas bloat, if you can get past the blockage. The blockage may impede the tube, and it is important not to force its passage. If you are unable to relieve the gas this way, contact the veterinarian urgently. They may need to trocharize the rumen.

Frothy bloat is more common. In this case, overactive microbes produce a foamy slime that coats the gas and seals it in the rumen. It usually happens when a goat eats a lot of food that it's unaccustomed to, which is why it's important to make changes to your goats' diet gradually. Passing a stomach tube will not free the gas, but it will allow you to introduce a product to break down the foam, which will help release the gas. If the tube alone provides relief, the bloat was due to free gas. Otherwise, you'll need to introduce a specialized goat bloat medicine from your veterinarian, typically poloxalene. If bloat is due to grain consumption, your veterinarian may provide an alcohol ethoxylate detergent as a more effective agent.

However, you need to act quickly, so if you do not have a veterinary product to hand, vegetable or mineral oil can be effective, although slower acting. Introduce 100–200 cc via the tube. Do not use linseed oil as it causes indigestion. If you cannot use a tube, you'll need to find someone who can. Using a drench without a tube risks product being breathed into the lungs and causing pneumonia. If this is your only option, take utmost care to avoid this risk. If you have no experience using a stomach tube, it's always best to contact your vet.

Hoof Rot

If you goats spent a lot of time on wet ground, during the winter, it can soften the hoof and make them more prone to bacterial infections. It's good to try and keep your pasture areas as dry as possible, and regularly check your goats' hooves. If your goat appears lame, or there is swelling between claws and a raised temperature, then they may be experiencing hoof rot. It will have a very unpleasant smell. You need to clean all the areas of the hoof you can, and trim off any areas that are rotten, it needs to be cleaned thoroughly with a treatment that has a tetracycline antibiotic in it. It can be hard to treat, you may regularly need to bathe the hoof and have veterinary prescribed antibiotics, and the hoof to be trimmed aggressively by a vet. You can also put down hoof rot mats if any of your goats appear to have this.

Johne's Disease

This disease also goes by the name of paratuberculosis. It's a fatal gastrointestinal disease caused by the bacterium Mycobacterium avium subspecies paratuberculosis (MAP). The bacteria incubate in the animal's body and can be spread amongst your goat herd and other animals, such as cows and sheep. The two main symptoms of Johne's disease are rapid weight loss and diarrhea. Goats may not always show the diarrhea. Goats may appear to have a healthy appetite, but still become thin and weak.

You will need a vet to come and send off samples to a lab to check for Johne's Disease. An animal with Johne's Disease won't be able to absorb the nutrition it needs from its intestines. The organism for it will be swallowed, may via contaminated water or milk. It is a disease that occurs in milk and meat goats. Ensure any new animals you bring into the herd have been tested for this. Repeat diagnostic tests numerous times, whilst the animal is quarantined from your other animals to prevent spreading this. You can do PCR tests on fecal samples, or tests on blood or milk. If adults are identified with this, they should be culled to ensure that no kids have contact with its manure or milk.

Mastitis

This is where a goat has udder damage. If conditions are not as sanitary as they should be, this can increase the chance of mastitis. One way to help prevent mastitis is to keep milking and living areas clean. You can see that a goat has mastitis if you look at the udder or look at the milk it produces, which may be abnormal consistency or color. The udder itself may be warm, swollen, and painful. Care needs to be taken to ensure that it doesn't become septic, if the goat has a fever, becomes thin or lethargic, these are all signs of Mastitis. If you see any of these signs, it's time to call your vet. The most reliable test for diagnosis is a microbiological culture, which can determine the cause of the infection through only one milk sample.

The treatment will be based on the microbiological milk culture, and can include antibiotics, ointments, corticosteroids, dry-off treatment and supportive care.

Parasitic Diseases

Parasites thrive better where it is cool and damp, rather than in the heat, so it may depend on the location you live in, as to how much bother these will be. You can have the fecal matter of goats checked for:

- **Barber Pole (Haemonchus):** This is a blood-sucking parasite that pierces the lining of the goat's stomach, it can cause anemia and death. Some goats have black or grey gums, which can make it difficult to see if their gums are lighter, but you can also look at the lower eyelid of a goat. If a goat has fluid accumulated under its chin, this is another sign of this disease.

- **Lungworms:** These are passed through feces and ingested by grazing animals. The parasites then go to the lungs and trachea. Young animals can become very weak and sick, and it can be fatal. Older animals may develop a fever, cough, and nasal discharge.

- **Tapeworms:** The most common tapeworm in goats is *Moniezia expansa*. Adult tapeworms live in the small intestine. They can cause intestinal blockages, but this is very rare. Typically, they simply rob your goat of nutrients and make it harder for them to gain weight.

Keeping your goats dewormed will help to prevent these parasites from attacking your goats.

Pink Eye

Goats can get this, just like humans can. Pink eye can be treated with antibiotics that are injected into the body or placed directly in the eye. The most common treatment is to apply terramycin ointment to the affected eye(s) 2–4 times per day. Keep the infected goat away from the herd and ensure you wash your hands very well after treating a goat for pink eye.

Sore Mouth/Contagious Ecythema/Orf

If your goat has lesions or blisters, or scabs on the mouth, you need to get your vet to check with this. You need to ensure your goat doesn't get any other bacterial infections in these sores. If you spot a goat displaying any of these, do isolate it immediately. Again, it is worth regularly opening your goats' mouths and taking a look to check they're healthy. Sore mouth is something which is contagious to humans too; if you think your goats have it, then anyone handling the goats must wear gloves. You don't want kids transmitting this to does' udders or to the assistant's hands. Another frequently heard term for this is 'Orf'.

Tuberculosis (TB)

Goats are fairly resistant to Tuberculosis, but unfortunately, they can still get it. Goats can have a skin test to diagnose whether they have this. Nodules form in the lymph nodes called "tubercles". It's a bacterial disease, causing illness, pneumonia, weight loss and death. The disease is passed through the air when infected goats cough near other goats, it can also be passed on by kids drinking infected milk. It is possible for a goat to pass on the bacteria throughout the herd, without showing any symptoms itself at that point. Signs of TB can include weakness, loss of appetite, weight loss, a fever that fluctuates, a bad cough, pneumonia, diarrhea, enlarged lymph nodes.

Upper Respiratory Infections

If your goat appears to have nasal discharge, coughs and sneezes a lot, and has a loss of appetite and high temperature, then you do need to contact a vet. These can be signs that your goat has an upper respiratory infection, which can impact the goats' lungs, nose, trachea, windpipe, and bronchi.

Urinary Calculi

If goats don't have a balance of goat minerals, especially phosphorus, this can cause wethers to have urinary calculi. It's like a kidney stone in humans, and like humans, these stones can sometime get stuck in the goat's urethra. Diet can contribute to this, so keep male goats off grain, and you may need to change the calcium or phosphorous ratio in your goat's diet, you could discuss this with the vet or a goat nutritionist.

How to Tell If Your Goat Is Sick

One way to tell if your goat is sick is based upon its weight, and whether it has lost weight. Here are some typical weights below of what types of goats should typically weigh:

- An adult doe from breeds such as Alpine, Nubian, Saanen and LaMancha would usually weigh between 125 lb and 175 lb.
- An adult buck or wether would usually weigh between 150 lb and 225 lb, from again Alpine, Nubian, Saanen and LaMancha breeds.
- Pygmy goats weigh approximately 40–80 lb for a female, and 60–90 lb for a male.

Another way to check your goat's health is to take its temperature, which typically should be between 101°F and 102°F.

If you have just purchased your goat, you need to ensure that they haven't become stressed in the shipping and got shipping fever, which can be seen with pneumonia, diarrhea, a fever of 105°F, nasal discharge, coughing and rapid breathing. If you think your goats have this, do contact an emergency vet.

Other signs that a goat is not well can include:

- Weakness or lethargy—if the goat is sitting, laying or still, it could be ill.
- Limping or staggering.
- Showing no interest in food and water.

- Soreness in the goats' mouth (do wash your hands if you spot this, as it can pass to humans).
- Pressing its head against a fence or wall.
- Ears held in an odd manner.
- Not urinating (it could have a stone blocking it).
- If the feces are abnormal (diarrhea).
- Pale eyelids and gums (if they are pale it could be anemia).
- If a goat is isolated and has wandered off alone, this can be a sign of illness.
- If the goat makes unusual vocalizations that are different from its usual bleats, it may be in pain.
- Check that the goat's tummy is not swollen or bloated and contact the vet immediately if it is.
- If the goat moves stiffly or has a hunched posture, it may be ill.
- If the goat shivers or trembles again contact the vet.
- If the goat's coat looks dull.

Goat First Aid Kit or Medical Box

Many goat owners would suggest that you put together a goat First Aid Kit, or Medical Box to treat common complaints at home, without needing to go to a farm store or vets. This is by no means an exhaustive list, it's what we have and use.

Items you could include are as follows:

First Aid Supplies

- 7% Iodine Solution or spray
- Antibiotic eye ointment
- Blood stop powder: it's always a good thing to have on hand.
- Blu-Kote: it's great to spray on wounds to prevent flies.

- Co-flex bandages: these can help you keep things stable if your goats break a leg on the weekend so that you can get them in the next day and avoid the monstrous out-of-hours charges.
- Disposable gloves
- Drench Gun and Tubing: 1–1½ inch works well for adult goats
- Headlamp or flashlight
- Heat lamp, heat pad, or other heat source: for warming a chilled goat
- Mastitis test strips
- Maxi pads
- Needles (22 g, 20 g, and 18 g)
- Notepad and pen
- Red top tubes: use for blood collection for mail-in tests like CAE & CL
- Scalpels
- Scissors
- Syringes (3 cc, 6 cc, 10 cc, 20 cc)
- Thermometer: it's important to know when a goat has a fever.
- Vet phone number: always have it at hand.
- Wound spray: it's good to have some antiseptic spray at hand.

Meds and Supplements

- Activated charcoal paste: neutralizes toxins.
- Baking soda: can help with bloat, and you can even leave it in the barn in a separate container. Goats are smart, and will take it when they need it to aid digestion, and leave it alone when they don't.
- Banamine: prescription painkiller. It may be rather expensive, but it's totally worth the cost.

- BO-SE: it's a selenium supplement. Great in case of floppy kid syndrome, or other symptoms of selenium deficiency. Many people give newborn kids a shot at birth anyways. It's prescription only, so you'll need to get it in advance from a vet.
- C&D antitoxin: most goat owners I know don't keep this, but we keep a few bottles on hand in the fridge. It helped us save one of our goats. If you suspect your goat might have entero—hit it with a C&D antitoxin.
- Di-Methox 40%: this is our go-to for coccidiosis prevention and cure. This will pretty much do the job in all but the most severe cases.
- Ferrodex injectable: injectable iron. If your goats have anemia, this will help them fight it and get back on their feet.
- Goat Nutri-drench: this is good to give your goats a boost if they're not well. We often give it to our does right after kidding.
- Lactated ringers: these are prescription only from the vet and they are used to prevent dehydration. When a goat is not eating or drinking, you need to keep it hydrated. Then you have a far better chance of fixing whatever is wrong with it because dehydration kills faster than most people expect.
- Maalox: this is great for an upset rumen, and in case of toxicity, it helps with diarrhea too.
- Molasses: old, plain molasses can be a great pick-me-up for goats who just kidded or are just chilly. Add a ¼ cup of this to two gallons of warm water. It will help to warm your goats up from the inside out, while giving essential sugars to help recovery at the same time. Our goats absolutely love this
- PenG: this is a penicillin antibiotic, it works really well on wounds, abscesses, and upper respiratory infections.
- Probios paste: goats' rumens need good bacteria to keep them running properly. This probiotic helps.

- Vitamin B complex: another great pick-me-up, especially useful when your goats don't have appetite or recovering.

Key takeaways from this chapter:

1. Take care to regularly monitor your goats and learn what is normal for them as a baseline, then monitor their health against that.
2. Keep health records for your goats.
3. Quarantine new goats to avoid spreading disease.
4. Ensure your goats have relevant vaccinations.
5. Avoid feeding goats grain if possible, especially male goats.
6. Be aware of various diseases that are common to goats so that you can easily spot the symptoms and can contact a vet at the first signs.
7. Be aware of common signs of sickness to look out for in goats, to give you an indication if your goat is not healthy.

The next chapter of this book will move on to look at breeding goats, looking after pregnant goats, births, newborn check, and cares, and raising kids.

Chapter 8: Breeding Goats

For most people I know, their goats having kids is one of their favorite times of the year. It is incredible to see new lives come into the world and fill your smallholding or farm with new energetic, cute lives. There is, however, a lot that needs to be done in preparation even prior to breeding starting, and therefore this chapter will cover preparing for breeding, caring for pregnant goats, birth, newborn checks, and raising kids.

Breeding seasons (known as rut) typically last from the end of July until the end of January. Most goats are seasonal breeders, and they start breeding when there is decreased daylight. It is possible to alter the breeding season by keeping animals indoors and using artificial lighting, but my personal preference with our goats is to keep things as natural as possible.

Female goats, which are termed does or nannies, usually give birth to one or two baby goats (known as kids). The gestation period (the length of time a goat is pregnant for) is usually 150–180 days.

Preparing for Breeding

Does need to keep a good body condition throughout the breeding season, they need to have excellent nutrition so that they produce healthy babies. If you are planning to produce meat from your goats, then it can be important to maintain a good body condition in the bucks too. It's worth paying more for a good buck, the best that you can afford.

It is possible to use a teaser buck on does. A teaser buck has been castrated and won't be able to make the does pregnant, but they can make the does go into heat. Some signs that a doe is in heat can include:

- Seeking out the buck and paying attention to him
- Eating less
- Behaving in a restless manner
- Peeing more frequently
- Making more vocal noises
- Mounting or mounting females
- A swollen vulva with mucus

Although goats can come into puberty as early as 4 months old, it is not advisable to breed them this young. As a rule, a goat should not be mated until it is one-year-old. Does may be mated when 10 to 15 months old so that they kid at the age of 15 to 20 months.

Preparing for breeding can be done a number of ways. It is possible to hand select the doe and buck, but this can be quite time consuming. Another option is to place a buck into a pen full of does and allow the buck to work out which does are in heat. This is less time consuming initially, but it can be hard to determine exact breeding record days and may need ultrasounds to work out when does are expecting. At 1 year of age, the buck should service no more than 10 does at a time (in one month). When he is 2 years old, he should be able to service 25 does at a time. At the age of 3 and older, he can breed up to 40 does at one time, as long as his health and nutritional needs are met.

Another option for breeding is Artificial Insemination (AI), but this can be costly, and it can be tricky to find people with AI experience, sometimes it can be a farm manager who takes on this role. It is also possible to use out-of-season breeding as previously mentioned using artificial lighting with goats that are housed indoors, it can mean that 60% more does can be used for breeding out of season. It is worth keeping good reproductive records for your goats, so as to prevent or reduce abortions, if does do not birth on the correct date after being identified as pregnant, or if there are abortions, or stillborns these are things that indicate there could be issues.

Chlamydia was covered in the previous chapter; however, I think it's important to repeat it here again. Chlamydial infection can cause abortions in goats. Chlamydiosis or *Chlamydophila abortus*, is a disease without symptoms and unfortunately, it's virtually undetected in a herd until multiple abortions occur. There is no general pre-breeding screening tool for does; however, it can be tested for in bucks' semen. It can be spread via reproductive fluids, aborted tissue of infected animals, and carrier animals born to infected animals. Pasture and bedding can also be contaminated and remain so anywhere from a few weeks to a few months, depending on environmental conditions. Diagnosing Chlamydia in goats is done by laboratory testing of placental tissue. Blood tests are not reliable unless they are taken at the time of abortion and again at three weeks. Chlamydia can be treated with Tetracycline or Tylosin, or other effective antibiotics. You should consult your vet for an appropriate course of treatment. Control measures include practicing good sanitation and establishing an effective vaccination program. Buy replacement does and kids from reputable sources with no history of the condition. Be aware that Chlamydia is contagious to humans.

It can be a good idea to increase the amount of food that is offered to the breeding does one or two months before breeding, this term is known as 'flushing'. I recommend that you give does ½ pound of grain, per head every day to improve their condition and help them to ovulate. Ensure the doe isn't overweight though, because this can cause them to have an uncomfortable pregnancy. If your goat is a milk goat, it can be a good idea to have a little bit more weight on them because once they start milking, it can be hard to get them to increase their weight.

Breeding goats is absolutely essential in you have goats mainly for milk. In order to give milk, a goat must first get pregnant and have kids of her own. So, when the does aren't pregnant (they are called 'open'), this can mean that you have a goat without any financial return.

We milk the goats for ten months and then dry them off two months before they have kids again around their next birthday. This dry period is necessary for them to have enough energy to grow their kids.

Two to three weeks before you start the breeding season, the does should be dewormed, vaccinated with tetanus, have their hooves trimmed if needed, and give the does a vitamin E injection to help with ovulation. If you are planning to use Artificial Insemination, then a cycle should be planned. Trimming the hooves can be important, to ensure their feet grow correctly, and because they'll be heavier when they're pregnant with kids too, putting more pressure on their feet. Increased growth hormone production during pregnancy can make their hooves grow faster too.

If you're placing a buck with does, you could place him with the does for 42–45 days and then remove him. If you have used Artificial Insemination and it hasn't got does pregnant, you could then opt to put a buck in with the does who didn't become pregnant from AI.

It can be a good idea to get the buck to wear a harness that contains colored crayon, because this will then allow you to know which does were bred from this buck.

As discussed earlier in Chapter 3, it is always important for goats to have shelter to keep them out of wind, rain, drafts, and sun or heat, but this is even more important when a goat is pregnant for five months.

A good piece of advice is that you should spend quite a lot of time handling your goats prior to getting them to breed because if you haven't, they can be tough to work with when they have pregnancy hormones and they're trying to care and protect their kids, so putting the time in to shape their behavior about being handled prior to this is a good use of time.

Caring for Pregnant Goats

Keep your does really calm and as stress free as possible in the three weeks after breeding, because this is when the pregnancy is starting, try not to change her daily routine, and try not to travel with her.

The gestation period for a goat is 150 days. It is sensible to perform and ultrasound on goats around the 32nd day of pregnancy. Some goats will have abortions. You can also confirm your doe's pregnancy with a blood test sent to a lab at 30 days, and if you chose to, you could run a check for CAE at the same time, which is mainly passed through the mother's milk, so it's good to check your goats are clear of it and won't pass it onto their kids (See Chapter 7 for more info on CAE).

If your pregnant goat is a milk goat, you can't milk her in the last two months of her pregnancy, she should be dry, because she needs all her energy towards creating healthy kids. It can be a time to treat does with an intra-mammary infusion to prevent mastitis, but do remember that there are milk and meat withdrawal times after using this medication.

In the last six weeks of pregnancy give a booster vaccination for the does of at least Clostridium, C&D, and tetanus so that the antibodies are passed on to the kids and will boost their immunity. Does can have increased feed if they're not up to the body condition they should be. It's recommended 5 for meat goats, and 3 for dairy goats. Goats' body condition is measured on a scale from 1 to 9, where 1 means emaciated and 9 is obese. 70% of the kid's fetus growth occurs in the last 6 weeks, so it is important. It's important for pregnant does to have high quality hay or alfalfa, and free choice minerals. You can give a second Vitamin E injection to help the embryo develop, and really gently handle does at this time so that they are not stressed.

If a doe is in a good body condition before giving birth, she should produce healthy kids that grow well and have good stamina.

You can introduce grain slowly to the doe one month before she gives birth because she will need those extra calories for milk production. Don't add the grain too fast though as you don't want to cause goat bloat (see Chapter 7 for more info on this).

You can also trim any long hairs around the doe's tail and back of legs about a month before she gives birth, this will make cleaning up the doe after birth easier.

Between 8 to 12 hours prior to giving birth to a kid, the doe's udders will develop, and the area around her vulva will become looser. The doe will lay down and stand up quite a few times. Does should have a calm, clean and dry place in which to give birth.

Birthing Kids

Ensure that the doe is giving birth in a clean, comfortable, and hygienic environment, fresh straw bedding can be better than woodchips for birth, because woodchips can be inhaled and stick on wet newborn kids. Having a clean and hygienic stall will minimize infection. Have your pregnant does in a private stall, which will be less stressful and less chaotic than in a herd environment, you can have pregnant does in their own stalls, but next to one another for company, or provide a goat companion.

Make sure that you have a vet's number at hand, or at the very least, someone who has experience in birthing goats. Have some powdered of frozen CAE-negative colostrum to feed the kids and disinfected feeding bottles. Have heating lamps, towels, disinfectant soap, water, lubricants, gloves, iodine naval dip, tube feeder, scissors, injectable vitamin E, syringes, and needles at hand.

The doe should be mostly left alone throughout giving birth, and giving birth to a kid should happen within 2 hours after the water sac has appeared. Kids will come out headfirst, leading with their front feet. Occasionally, if it is a more difficult birth, then they can lead with their hind legs. Do be prepared to help with the kidding, ensure your nails are trimmed and clean, and use gloves. If your doe is struggling to give birth, then it is advisable to call your vet.

Sometimes your goats may need cesarean sections, it's good to be fully briefed on goat labor so that you know what to look for. If you need an emergency visit to the vets with a goat, it is likely to cost over $100. If a goat needs a cesarean, it will be over $500 and if you need a vet to come out to the farm for an out of hours visit, this is likely to be $800 or more.

Newborn Check and Care

Ensure that newborn kids are breathing, remove any material from their mouth and nose with a clean towel. Keep the newborns warm and dry, if bedding becomes wet, change it, and put fresh bedding down. If the weather is cold, you can wrap the kids in a towel and use a heating lamp to warm up the temperature (but make sure this is safely installed so as to not cause injuries, and that the cord cannot be chewed). The kids' umbilical cord should be cut to 1.5 inches, and then sprayed with 7% iodine to reduce infection. The kids should be kept with the does unless they are CAE positive, or you are specifically bottle feeding them.

The newborn kids should stand up very quickly after being born. Ideally, kids should be given a good amount of colostrum as soon as possible after birth, definitely within the first eight hours. If the kids are weak, help them to stand and suckle. Bottle feed colostrum to kids if the does don't produce enough milk, or if they can't stand up and suckle. Colostrum can be warmed to 100°F–102°F (38°C–39°C) before feeding. You want to ensure kids have had enough milk, but without forcing them to drink, as that can cause diarrhea.

In my experience, it is definitely worth looking out for newborn kids and their mothers and support where needed. Here are the things you should look out for:

- Adverse weather conditions of cold/rain—kids can lose energy and die.
- Kids being abandoned by their mothers.
- Poorly does, who are sick and don't pay attention or can't provide nutrition for the kids.
- In case of a multiple birth, if a doe has three or more kids, she may not be able to care and nourish all of them. What can be done, is that if another goat has a single birth, you could wash a goat from the multiple birth in the single goat's amniotic fluid, and give it to the single goat's mum so she thinks she's had 2 kids. She will usually clean the newborn and care for it as her own.
- Weak newborns who are unable to stand or suckle.
- Predators, keep the area around the birthing stalls and newborns highly protected to avoid any newborns being killed or injured.
- Ensure that none of the herd are being aggressive towards the newborn kids. Sometimes if a kid got into another pen by accident, the doe in there may try to butt it.

If the kids are raised with the mother, then they will feed off her milk a few times a day. If they are being weaned, then the goat kid should be fed approximately four times a day for the first few days, and then you can move to feeding them twice daily.

Raising Kids

Does will usually take good care of their kids, and you will need to give minimal attention. Does will lick their kids clean after birth and may bleat to the kids to attract their attention.

Kids will get up and walk around within a few minutes of being born and should suckle about 30 mins or so after birth.

Does will protect their kids from other goats in the herd and will feed and nourish their offspring to help them grow and develop. By 3–4 months, the kid will be weaned (no longer need to nurse from its mother).

Kids can start to be weaned naturally or off milk when a kid has reached between six to eight weeks, and this will depend on how much forage and water the kids eat. You can offer forage from 2 weeks onwards.

As mentioned previously, goats can become fertile by the time they have reached 4 months old, so it can be important to separate does and bucks when they are young. When they have reached 70 percent of their adult body weight, they can breed, but if you breed too early, it can mean they don't produce as much milk, or live as long. It's not a good idea to overfeed them to try to increase their body weight either. My advice here again would be for the best health, let things be as natural as possible, and don't force things unnaturally.

When we first bred our does, it was wonderful to see the tiny little kids being born, and the care their mothers took of them. It's an amazing process, from start to finish. We did have a scenario where one of our does had three kids, and we decided to bathe one of her does in another goat's amniotic fluid who had just had one kid, we placed the goat in a bucket with the fluid and ensured it was covered, then placed it with the singleton pregnancy, and she took to the kid, just as though it were hers and the kid started to suckle from her. The other doe had two other kids, which were more than enough for her to deal with and to provide nutrition for. All the goats remain in the herd and seem very happy. Goats are enormous fun and goat kids are just hilarious and so cute!

Key takeaways from this chapter:
1. Before breeding, handle you doe regularly so that she is used to this.
2. Trim your does' hooves during pregnancy.
3. Ensure that you doe is a healthy weight, and give her high quality hay/alfalfa, in the last month of pregnancy, you can slowly add some grain. Don't forget the minerals too!
4. You can confirm pregnancy with an ultrasound or blood test.
5. Do a CAE test when pregnant so that mothers can't pass this onto kids through milk.
6. Ensure the doe is dry (not being milked) for the last two months of pregnancy.
7. Make sure that throughout pregnancy your doe has good shelter with no drafts, and before birth that she has her own individual pen, with goat friends nearby for company.
8. Make sure pregnant goats have had the CD&T vaccine so that they will pass on immunity to their kids.
9. Have a clean stall for the doe to give birth in.
10. Have plenty of powdered or frozen colostrum in.
11. Have contact numbers of vets or goat mentors.
12. Watch for changes in goat behavior that show birth is imminent.

The next chapter of this book moves on to looking at keeping goats specifically for milking, and what you need to know, preparation before milking, the milking procedure, how to handle milk to sell it to humans, and about the equipment you require for this.

Chapter 9: Milking

Can you think of anything better than producing lovely fresh goat's milk that you can turn into a wealth of amazing products, such as cheese, yoghurt, kefir, ice cream, as well as toiletry products. Goat's milk is excellent for people who have allergies to cow's milk, so it provides a great alternative. It also makes lovely soap that is very calming and gentle on the skin.

Some people think that goat's milk may taste musky or have an odor, but it doesn't! It is sweet, clean, and fresh. You will find it tastes just like cow's milk that most people are used to.

This chapter will cover the essential information you need to know about milking goats, from how to best prepare to milk, the milking procedure, handling milk to ensure it's fit for human consumption, the equipment you'll require, and how best to clean it.

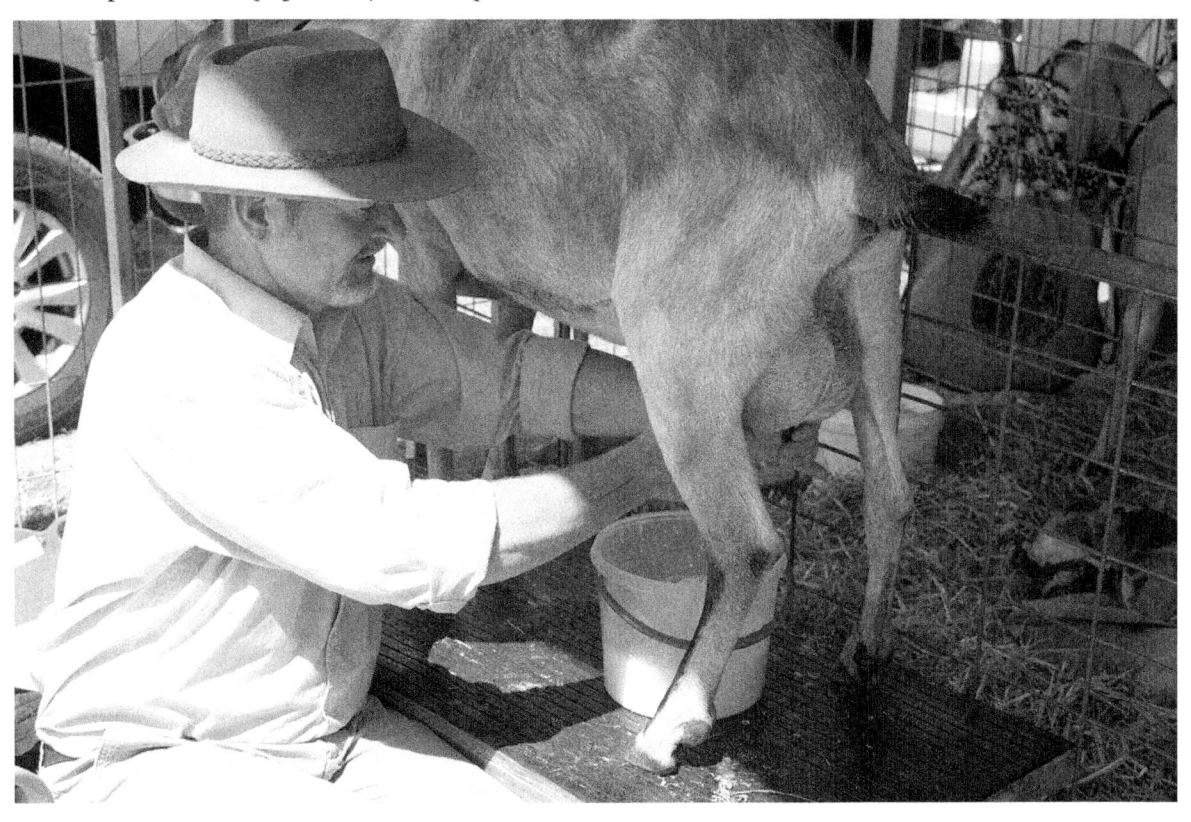

Milking Essentials

If you want to milk a doe, I would highly recommend that you get a milking stand. No doe is just willing to stand still and let you milk it. You will also need a milking pail (a stainless-steel pot); a cheesecloth (which can be boiled twice a day after use) or disposable milk filters; and a strainer (which can be boiled again). If you have bought disposable filters, simply discard them after use, otherwise they will smell unpleasant. If you have two to three goats that you are milking by hand, a 4-quart stainless steel pail should be sufficient for you.

Some people choose to weigh the food they give their goats, and also to weigh the amount of milk produced too. You can keep records too and I highly recommend that you do that.

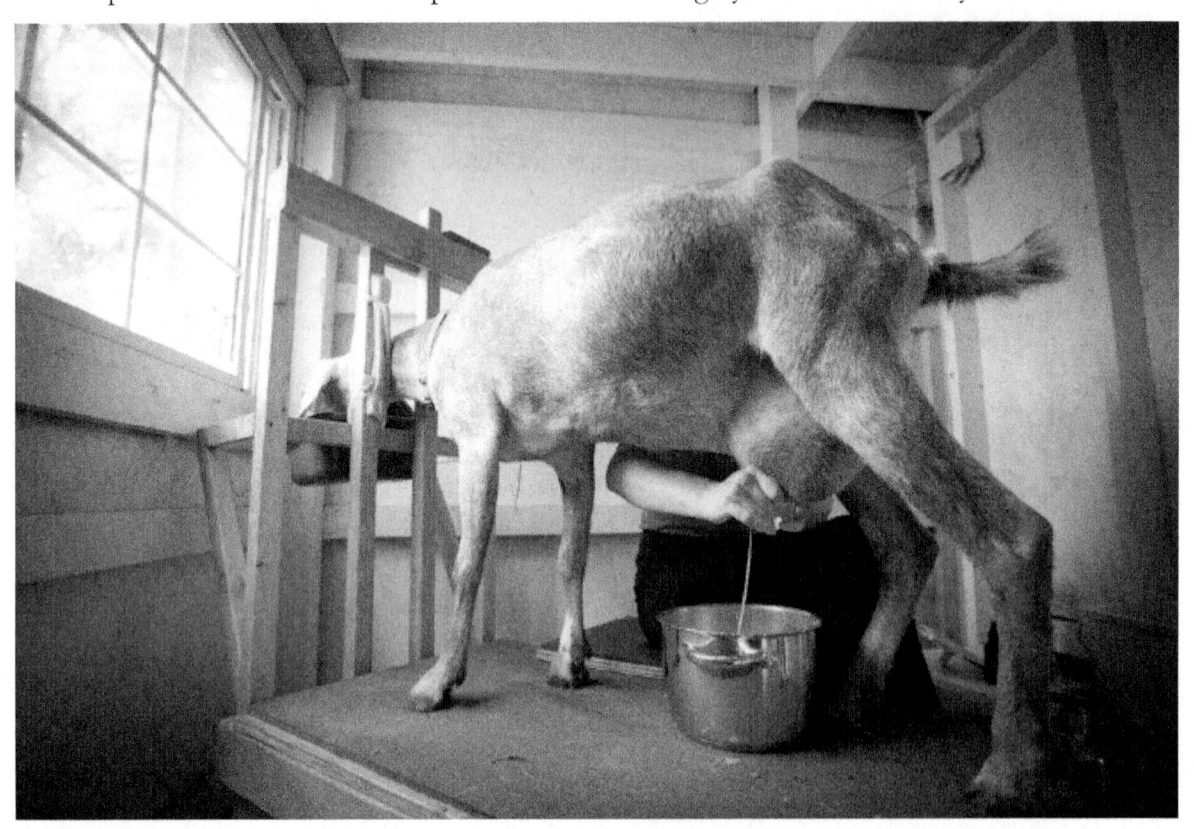

Preparing to Milk

It's good to have a cloth which has been soaked in water, with the smallest amount of bleach diluted in it, to clean the goat's teats, legs, thighs, and buckets before milking, because this will minimize any hair or dirt that will fall into the milking pail bucket. It can be good to keep the goat's udder shaved, because this will make milking much easier.

Even before you get to the milking stage, it is best to keep bucks away from your does for milking, because goats are sensitive to pheromones, and the milk may taste musky if a male buck is giving off his scent near the females (except during breeding).

Before you start the proper milking, it's good to squirt each teat, as this will get rid of any blockages in the teat and bacteria. Check this first squirt to make sure the milk does not contain blood or lumps of milk (if it does, this could mean that your goat has mastitis and needs to be checked by a vet).

Milking Procedure

It's good to find a milking schedule that works for you. Goats should be milked twice a day. Whilst many people say that milking should have 12 hours in between, sometimes that doesn't fit in with real life, so some people may opt to milk at 7am and 4:30pm, for example. There isn't quite 12 hours in between, and this may mean that you produce more milk in the morning and less in the afternoon, but your milking schedule has to be convenient too.

Goats can be a bit stubborn, and you may initially hold the goat's leg with your left hand, but with time, you will learn when your goat is about to kick, you need to be strong when this is happening.

A really sound bit of advice is to prepare some treats for your goat whilst you are milking her. You may choose to use some Alfalfa or Bermuda pellets and some organic grain, or alfalfa hay or sprouted barley grass.

In order to milk the goat, hold your index finger and thumb together like in an "OK" sign and use this to hold the teat firmly, then use the other fingers to press the teat to your palm and this pressure is what should squirt the milk out. You need to trap the milk and squeeze it out, trap it and squeeze it. It can be a bit tricky in the beginning, but with practice, you can learn to do this with both hands at the same time. If you don't get any milk, or only a thin stream, it can mean that you haven't kept your thumb and index finger pinched together enough. When you're milking your goat, it isn't about pulling or tugging down on the udder, it's instead about pinching and squeezing.

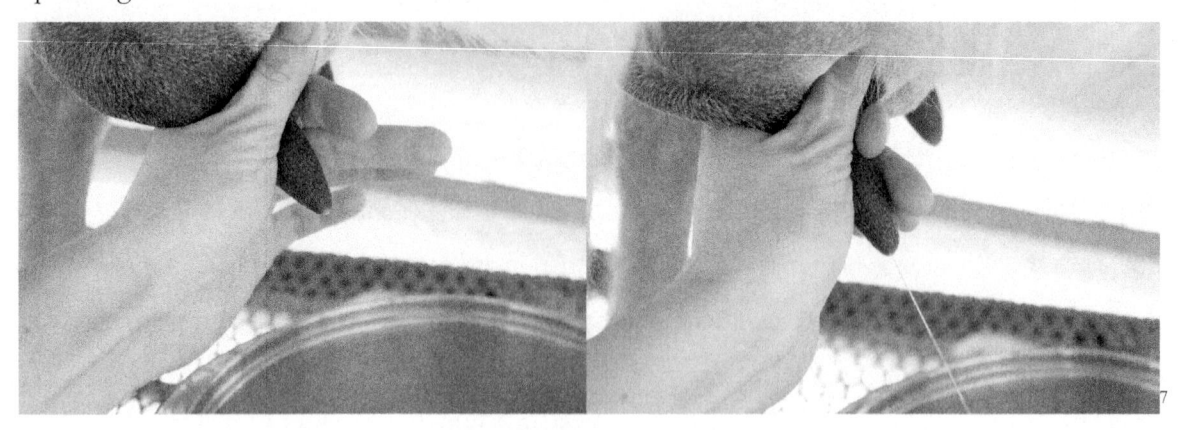

If the udder is full, it can be hard to get the milk to go in the direction you want it to, but when the udder empties a bit, it's easier. As the udder empties you need to massage it from top to bottom, to get the milk down into the teats to trap the milk and squeeze it out. It's really important that milk the goat completely and get all the milk out, this is super important, because if you don't, it can lead to mastitis and infection. You do need to massage the teat thoroughly to ensure there is no more milk in there. Another movement you can do, is when you think all the milk has come out very gently nudge your fist up into the udder, as though your fist was a baby goats head nudging the udder upwards, which can release and stimulate any other milk in there to come out. When the udder is all done, it may have a wrinkled appearance. You can then rub udder balm over the udder and teats to keep them healthy.

Pasteurizing Goat's Milk and Keeping It Fresh

Raw milk is something which can perish quickly. It is best that everything you use is stainless steel or sterilized glass.

---

[7] Images from weedemandreap.com: https://www.weedemandreap.com/how-to-milk-a-goat/

If you have kept bucks away from your does, there is no reason why your milk should taste musky. The only other thing you need to be careful with to ensure it doesn't taste "off" is the way that you handle it.

So, here are some useful tips to help keep raw goat's milk really fresh.

When goats are milked, they are milked directly into a clean milking pail (a stainless-steel pot or bucket. If you have pygmy or dwarf goats, you may need a shorter bucket than with full sized goats, in order to fit under them). This milk can first be weighed so that you know how much milk each goat is producing.

Then the next stage is to filter the milk. To do this, it can be passed through a cheesecloth-lined strainer, which will ensure that any bits of goat hair, dirt or bugs that have fallen into the milk are sifted out. It's not too bad, and don't let reading this put you off, but it's just to keep things super hygienic and clean. If you don't want to use cheesecloth, you can also purchase disposable milk filters. This then needs covering to keep out dust and flies, and this raw milk is then good.

The milk should be chilled as soon as possible, definitely within 30 mins of milking. If you have a fridge that can cool down to 35°F–38°F, then milk can last up to 10 days. However, most fridges only go down to 40°F, so in that case milk will last up to 5 days. If you put the milk into jars, leave an air pocket so that if you decide to freeze it, it won't break the jar.

Initially when we were very small scale, we bought a mini-fridge especially for goat's milk that would hold about 4 gallons of milk, but now we've scaled up and make cheese, yoghurt, and ice-cream from more dairy goats, so we have industrial commercial fridges.

It is best to store milk in glass jars; you can buy half-gallon mason jars with plastic lids.

If you intend to turn the milk into cheese, have a tracking system that records the date of the milk on each glass jar. Don't leave the glass jars in sunlight or fluorescent light as it can change the flavor. If the milk begins to taste of "goat" as it ages, note that this taste would be passed on to anything like cheese too if you were to make cheese out of it.

If you're intending to make cheese or dulce de leche out of the goat's milk, you should definitely do this within 4 days of milking the goat. If you intend to make yogurt of chevre, that should be done the same day, or next day after milking.

How to Pasteurize Goat's Milk

It's up to you whether you want to drink raw or pasteurized milk. I am certainly no medical professional, so I will only present both sides of the argument.

Pasteurizing milk kills all bacteria, both good and bad, so some people prefer to drink raw milk as it has good bacteria which aid gut health and improve digestion. However, raw milk may contain bad bacteria which can make you sick. Pasteurized milk lasts longer and is safer to drink. With that said, here is how you can pasteurize goat's milk.

There are three ways to do it: using a milk pasteurizer, double boiler or simply on the stove.

1. **Milk pasteurizer**

Using a milk pasteurizer makes the whole process super easy. Simply pour your filtered milk into a stainless-steel container and put it inside the heating mechanism, which works as a double boiler system. It heats the milk to 165°F for 15 seconds. This is all the time and heat it takes to make milk safer for drinking.

When it's done, remove the container and place it in an ice bath to cool it down quickly. This gives the milk a fresher taste. When milk has cooled to approximately 55°F, you can bottle it in sterilized jars and store it in the fridge.

The only downside to this method is that a milk pasteurizer does cost quite a bit, so if you don't want to invest hundreds of dollars into a milk pasteurizer, you can make your own double boiler.

2. **Double boiler**

To make your own double boiler, simply place a stainless-steel bowl over a pot of boiling water. Add the milk to the bowl above the water and stir frequently to avoid scorching. Heat the milk until it has reached 165°F for 15 seconds. Then, cool the milk off in an ice bath. Be sure to use a thermometer to measure the temperature of the milk and don't touch the pan, as this can cause the reading to be inaccurate. When milk has cooled to approximately 55°F, bottle it in sterilized jars and store it in the fridge.

3. **On the stove**

This is perhaps the easiest DIY method for pasteurizing milk at home. Pour the filtered milk into a stainless-steel pot and heat it on medium heat to 165°F for 15 seconds, just like in the two previous methods. Stir the milk frequently to avoid scorching it and use a thermometer to measure the temperature of the milk. Then put the pot into an ice bath. Once again, when milk has cooled to 55°F, bottle it in sterilized jars and store it in the fridge.

And that's it! Your milk is now pasteurized and safe to drink.

Cleaning Your Equipment

Rinse out the milk pail several times with cool water, spray it with a cleaner, and then use boiling water to disinfect anything that has come in contact with the milk. If you used hot water initially, it can make the milk stick to surfaces and become hard to remove. You would clean out glass mason jars that you've stored milk in, in a similar way, rinse out with cold water, spray with a cleaner, then boil in hot water or put them through a dishwasher. A few times a month you can use a solution of 2 tablespoons of 3% hydrogen peroxide inside jars and pails to coat insides of both, then put them in a dishwasher if they fit, or boil them in hot water to sterilize them.

When we first got our goats, we didn't have a milking stand, and it was like a scene from a comedy sketch as I tried to milk a doe who didn't want to be milked and tried to wander around while I was trying to milk it, and I got repeatedly kicked by her too! Once we got a milk stand, and with time, I became more confident at milking, it was a much smoother process all round. It is lovely to have fresh goat's milk and you can make so many fabulous dairy products from it. It's a sense of satisfaction that I can barely describe when you produce your first lot of cheese, and first lot of yoghurt, and ice cream. Knowing that you raised the animals that produced the milk to create this is an enormous sense of achievement and pride.

Key takeaways from this chapter:

1. Buy a milk stand, and stainless-steel pails and strainers, plus either cheesecloth or disposable filters.
2. Clean goats' teats before milking.
3. Get into a good routine about when you milk your goats (milking your goats every 12 hours is recommended but not always 100% practical).
4. Keep bucks away from does so that their musky pheromones can't influence the taste of the milk.
5. Provide goats with grain and good minerals to keep the goats as healthy as possible.
6. When you wash out milk items, wash the milk residue off with cold water a few times, then boil it or put it in a dishwasher.

7. Sterilize items with a mild bleach solution a few times a month.
8. When milking a goat, the most important thing is to massage the teat to ensure all the milk is out, keep massaging down. If there is milk left in, it can cause infection and mastitis.
9. After filtering the milk, get it chilled as swiftly as you possibly can, definitely within 30 minutes.
10. If you're intending to make cheese, chevre, yoghurt, or dulce de leche out of the goat's milk, do this as soon as you can too.

The next chapter of the book will move on to look at having goats for goat meat, and when and how to butcher a goat, how to determine the quality of meat, and about storing meat.

Chapter 10: Harvesting Goat Meat

Many people have been raised to eat chicken, pork and beef, and goat may not immediately jump into their mind. But many countries have goat as part of their menu for a good reason. Goats are versatile and easy to care for and require less land than cows. Goat meat has less fat than other types of red meat, and the fact it is lean can make it a popular choice for people to buy.

If you want to store meat from an animal, cows produce a lot of meat, whereas the meat from a goat is much more manageable to store.

A key tip for people raising goats in the US for meat is to check the zoning requirements prior to purchasing them because in different states goat definitions vary from 'livestock' to 'companion animals', and this could have an impact on how successful your business is.

Some breeds of goats bleat more loudly than others, so if you are in a suburban area where neighbors could complain about the noise, this is worth thinking about.

You need to consider whether there is a market to support the number of goats you have. Do you want to raise goats for meat for yourself or to sell? Will you be able to sell all culled animals? Will you breed goats to make replacement goats or buy new ones? You need to think about this before purchasing goats.

So, assuming you have done all the above research, and have goats, you may wonder at what stage of the goat's life should you butcher it for meat? The answers are below.

When to Butcher a Goat

It is generally bucks that are harvested because does tend to be kept for breeding and producing milk. A good age for a goat to be butchered is 8–10 months old, it's not just about when a goat is this high or weighs that many pounds, it's about how much the goat will yield. This will vary from breed to breed. To evaluate when your goat is ready, you will need to look at how much muscle it has on its forearm, and its overall appearance. The more muscle your goat has, the more meat you will get from it. A goat's forearm is the top of its front leg where the leg meets its body. Your goat will grow taller and bigger first, and then will develop muscle after that. If your goat doesn't have much muscle, allow it more time to grow. Another place that you can check the muscle of a goat is the outward facing muscle on the goat's rear right leg where the leg joins the body, this area is called the stifle, this part of the goat should have a definite outward bulge, when you are viewing the goat from behind. The goat should have muscle in its inner thighs that join the center of its body. If the goat doesn't have muscle, leave it longer to grow and develop. If you still have the doe that gave birth to the kid, compare the kid with her, and the kid should have more meat on it than the doe. You can also take pictures of the kids from time to time, at the front, rear, and side to give yourself something to compare it to, to see how it has grown. It can be easier to see this in photos because when you're looking at the kids twice a day at feeding time, you don't tend to notice the changes.

As a general rule of thumb, whatever the carcass of the goat is, you will yield about 50% of that in meat. So, if your goat weighs 80 lb, you'll get approximately 40 lb of meat.

The goat you butcher should be approximately 8–10 months old. Sometimes people want a 50–60 lb roaster goat, which is younger than 8 months.

If you want to get your goat to a point where it can be butchered sooner, you can feed it some grain, or pellet feed. Generally, goats that just have forage take longer to reach a weight where they can be butchered. If goats have been fed a concentrate diet, there were less off-flavors reported, compared to goats that were fed from the range.

Boers are a meat goat, and they will need supplemental food in order to grow to an optimal weight to be butchered. If you just want your goats to forage, then a better breed to choose is Kiko or Spanish.

Many people try to get a goat between 80 lb and 100 lb in weight, you can weigh your goat each month to see how they are progressing. If you don't have a livestock scale, you can lift the goat onto your normal bathroom scales in your arms, then deduct your own weight, to work out what the goat weighs.

If your goat isn't looking as well as you'd like it to, you need to ensure it has been dewormed and that it has more calories to eat. It may also simply need more time to develop muscles. You can't hurry this along too much, goats need time to grow.

How to Butcher a Goat

To butcher a goat, it is a similar technique to butchering a lamb or deer. Simply put, you'll gut the animal, remove the head, work the skin off, and rinse the carcass once this has been done.

If you are planning on butchering a goat, you can do this when a goat is approximately 8 months old. If you are butchering the goat yourself, you will need to slit its throat, but it is kinder to make the goat unconscious first. A goat's head is VERY hard, so you will need to be strong and use a very heavy implement. When the goat is stunned, you will be able to string it up, which will make the next stages easier. If you save the blood from the goat, it's recommended that you add a splash of red wine to it, and stir it thoroughly, this will prevent the blood from clotting into a lump. You can use the goat's blood in recipes, for example blood sausages, but you can also use it in rice dishes too. Some people, including many top chefs and restauranteurs, believe that if an animal is killed for food, it makes sense and honors the animal and its sacrifice of life, to use as much of that animal in every way you can so that it doesn't go to waste.

In order to skin the goat, start by cutting the skin at the back legs, then move downwards to separate the skin from the body of the goat. There will be membrane attaching it which can be loosened with a knife.

You can cut around the goat's anus and remove the genitals. The goat's testicles, otherwise known as "sweet meat" or "sweetbreads" are a delicacy which can be poached and then cooked in something like an omelet.

If all the skin has been loosened with a sharp knife, you should be able to pull it downwards to remove it, you do need to be strong and firm when you are pulling this. If you start to tear the meat, you need to loosen the skin some more with the knife. When you have done this, you can cut the goat's head off with a sharp butcher's axe. You can cook the goat's brains if you wish to do so, and again the skull can be split between the horns using the butcher's axe.

You need to gut the goat next, so start by cutting between the legs, down towards the rib cage, and have a bowl to catch all the innards that you pull out. You can tie the intestines so that they don't leak over your meat. You can put the liver and kidneys to one side to be used for cooking.

You can tear the diaphragm area to remove the heart and lungs and put them with the liver and kidneys you have set aside.

You will want to remove the intestine and any other tubes or bits of gristle. You can tie the top of the intestines to pull out. This will take a lot of work and strength to pull out.

You need to chop off the feet using the butcher's axe; then the goat carcass needs to be given a really good wash. It is best to hang the carcass after skinning for 24 hours, before you cut that up into relevant pieces.

If you are trying to utilize as much of the animal as you can, you can boil the bones to make soup.

You will want to remove tenderloins first, using a knife to start where the loins rest behind the ribs, then remove the rear legs, you can use a hack saw to cut through the spine. You next want the flank steaks, then the back strap, before getting the rib meat from the carcass. Then next is the shanks from the front legs and the shoulders, finally the neck meat can be used for stew.

I would, however, advise in most instances, you employ an abattoir or a butcher to perform this job for you, rather than doing it yourself, especially if you've never butchered an animal before. It will also allow you to watch and learn, and once you're confident enough, you can try to butcher your goats yourself.

Meat Quality

Research has shown that when people have goat meat from younger animals, the meat is tenderer, and doesn't taste "off" compared to older animals. If meat is from a dairy goat or a meat goat can make a real difference to the quality of meat too. Cabrito is meat from young milk-fed goats that weigh between 15 lb and 25 lb, and they are usually harvested at a month to two months old. People tend to barbecue this type of meat. Chevon is the name for meat coming from goats who are 6–9 months old, they have eaten forage and some grain and typically weigh 50–60 lb. Cabrito meat is more tender than Chevon, it has a similar tenderness to lamb. If a goat is larger than Chevon, this is still meat that can be sold. If goats have increased fat content of high protein grain for a long time, this can increase the aroma and flavor of their meat. Essentially, the older a goat is, the tougher the meat will be. Goats can be butchered up to 16 months, but this would be very rare for this to happen. Bucks are usually harvested before they reach puberty because then bucks take on a distinctive smell which can permeate their meat. Most young bucks are wethers that have been castrated.

Goat meat is lean and bright red and doesn't have fat marbling that you see in cows. Young goats have thicker muscles than older goats. If you are buying goat meat from a butcher, you want good, thick meaty shanks, not thin or narrow ones.

Storing Goat Meat

If you are intending on using the goat meat very soon, it can be refrigerated. If you want it for the longer term, then freezing is a better option. If you have small chunks of Chevon or it's ground meat, you can keep it in your fridge, but use it within two days. If you have whole cuts of meat in the fridge, they will be fine there for 3 to 5 days, but they need using at that point.

If goat meat has been properly packaged and frozen, it can be kept in the freezer for approximately 3–4 months, this is for ground or cubed Chevon. Larger cuts of meat can be kept frozen for a bit longer, up to 6 months, such as steaks, chops, or legs. If it is kept frozen for longer than this, it can taste off and be damaged by the freezer. It's good practice to label any freezer meat, and date it so that you know exactly how old it is.

When frozen goat meat is defrosted, it's best to thaw it in the refrigerator, it's the safest way to do it. Depending on the size of the meat, this can take 24 hours or longer. You want to ensure the goat meat is in a leak-proof container so that it does not leak any blood elsewhere and contaminate other food in your fridge or create a mess. You can also defrost meat in airtight packaging in cold water or use a microwave, but the fridge is the best option. Never re-freeze thawed meat because once the meat reaches a certain temperature, bacteria could multiply and make you sick.

I still remember the first time I tasted goat meat. To be perfectly honest, I was a bit hesitant to try it initially, because I love goats and because it was a different type of meat to beef, pork, lamb, or chicken that I'd eaten throughout my childhood, it was unknown to me. But it is beautifully tender, delicious, flavorsome, and much healthier than many other types of meat.

Key takeaways from this chapter:

1. Goat meat is very lean.
2. It's relatively easy to store goat meat because they are not huge animals.
3. It's more typical for bucks to be harvested for meat rather than does.
4. Typically, a buck should be butchered at 8–10 months.
5. Check how much muscle the goat has before butchering, if it does not have enough, give it more time to grow.
6. To butcher a goat, gut the animal, remove the head, work the skin off, and rinse the carcass. Or preferably get a butcher to do this for you.
7. You will need a butcher's axe to butcher a goat.
8. Younger goat meat is generally more tender, cabrito is harvested at a month to two months old. Chevon is goat meat from a goat 6–9 months old.
9. Goat meat can be refrigerated but needs eating quickly, 2 days for cubed and ground goat meat, and 3–5 days for whole cuts.
10. It is best to thaw frozen goat meat in a refrigerator.

As well as goats being fantastic for meat and dairy, there are many additional benefits of having goats, and this is what the next chapter will move on to discuss, looking at fiber, weed control, manure, using goats as pack animals and having them as pets.

Chapter 11: Additional Benefits of Goats

In earlier chapters, we have looked at how goats can be used for milk and meat. This chapter covers some other things that goats can be used for to make a small business and bring in some profit to help you live sustainably. These include getting fiber from goats, using goats to weed, using or selling goat manure, and having goats as pack animals. You can also have pet goats too, as they make wonderful companions.

Fiber

There are four key types of goat that can be used to produce fiber: Angora, Cashmere, Pygora, and Nigora. Cashmere wool will get you $120–$140 per pound. Typically, a cashmere goat will give you 6 ounces of wool per year. You need 16 ounces to make an adult sweater and approximately 1 ounce to make a scarf. So, you would need three goats to make a sweater, and you could make 6 scarves out of one goat's wool for a year. Cashmere wool comes from the undercoat of a Cashmere goat. Cashmere wool is lightweight, but incredibly warm, it is soft and durable.

Cashmere Goat

Angora Goat

Angora mohair gets $10 per pound if the mohair is carded (this is where fibers have been layered to transform it into yarn by spinning) or dyed the price increases. Mohair is fluffy and soft and used to make lace, knit, weave, socks, hats, sweaters, scarves, and gloves. Angora mohair is different from "Angora wool" from Angora rabbits. The mohair that is sheared during the "first clip" (the first shearing) is typically the softest, and this can often be used for baby blankets or clothes for newborns.

If you want to card wool yourself, you can buy a mohair carding brush, or you can use a horse or dog brush that will do the same job. Mohair is graded by character and style, and the style refers to how crimped the mohair is. You can buy spinning wheels that are small and portable, or larger scale spinners. You don't have to spin the wool yourself. You can sell raw mohair or cashmere. But, if you clean, dye, and spin your goat fleece, this will bring you in more income.

If you are purchasing goats for their fiber, then you will mostly want to buy ones with white fleece, because it can be dyed any shade. Some large fiber warehouses will not take fiber that is colored. But, if you are selling to places that value natural products and have brown, red, or black Angora fiber, this may get you $10–24 per pound. If you deliberately want to purchase a colored Angora goat, it is likely to cost $300 - $600 and a silver goat would be even more expensive. Cashmere goats similarly are preferred white so that their wool can be dyed. Cashmere fiber needs to be a minimum of one and one-quarter inch long, with a diameter of 19 microns in order to sell it. Most cashmere comes from China.

Pygora and Nigora goats are cross breeds. A Pygora is an Angora and Pygmy goat cross breed. A Nigora is an Angora and Nigerian Dwarf goat cross breed. Both of these give you miniature fiber goats. Pygora goat mohair can be dyed, but naturally comes in black, white, brown, gray, and camel. A Pygora goat usually lives for about 12–15 years, and their mohair remains soft and fluffy all throughout their life. Nigora goats produce fleece called "cashgora", a mix of mohair and cashmere. Pygora fleece is split into three categories. Type A is the most Angora looking with ringlets and sheen. Type B is a soft mix between Type A and Type C. Type C is that which is most like cashmere, it doesn't have any ringlets, and instead gives the goats a soft "aura" like appearance. Type A mohair has the least amount of guard hair, which is the outer coat of coarse hair. These hairs are removed by hand because it's too fine for a machine.

If you are shearing a goat for fiber, ensure that the goat has been deloused to get rid of parasites a few weeks before shearing. Wash the goat and with a cool hair dryer blow dry the coat to get rid of any dirt and debris. Allow the goat's coat to fully dry for 24 hours before shearing. Never shear a goat if its coat is wet, as the wool will get matted in the shears and cause pain and may cut the goat. It's beneficial to shear kids first, then older goats, to separate the first clip from the other fleece. The first clip will give you the most amount of money.

How to Shear a Goat

First, gather all the necessary supplies. Here's what you'll need: the clippers or scissors, lead and halter, and bags for collecting the fiber. I recommend using a goat stand or milking stand to shear your goats. You can have someone hold the animal, but a stand is still a much better choice. Restraining the goat's head helps keep it still. Just like when milking your goats, you can offer treats as you go to keep your goats happy. If a goat is always trying to kick you when you are shearing, you can put hobbles on their back legs, it helps tremendously. Be prepared before you start shearing your goats and always try to keep them restrained as little as possible.

Now, for the choice of your tool for shearing. You can shear your goats using either scissors or electric clippers. Spring-loaded scissors are a great tool for shearing your goats, but it can be time consuming. It still allows you to spend some time with each goat, they thrive on attention. Electric clippers, however, are much quicker and finish the job in much less time. They are more expensive and require regular maintenance, though.

When you're shearing a goat for fiber, start by shearing a strip along the backbone. Your first cut should be along the top line of the goat, from tail to shoulder area. Then shear either side down the sides of the goat, working from top to bottom, keeping blades parallel so that the animal is not nicked. If you're using clippers, use long, smooth swipes. Always try to complete long complete swipes first, and don't go back for cleanup swipes until you've sheared a goat completely. Avoid short clips, as they leave you with short unusable fiber. Follow along the lines of your goat's body and don't pull the fiber away to shear it, as it often leads to cuts.

Now pick up the fiber you've sheared and clean up any areas that still need to be trimmed. Collect the trimmed fiber and move on to the chest and the neck.

Not it's time to finish up the shearing. Belly, the britch area, and lower legs are not processed into roving or yarn most of the time. The fiber in these parts is often stained, felted, matted, or too short in length. Make sure to cut the fiber from armpits, goats can have armpit hair, unlike sheep, for example. Now trim off the fiber around the top of the head and ears. It's usually clean, but too short in length. But it can be used for stuffing things or used for doll hair. One important thing: Do NOT trim the hair at the end of the penis. It directs the stream of urine away from the body and it's necessary, so don't touch it.

When shaving under the goat, go slowly and have good visibility so that you don't get too close to the penis, teats, udders, or testicles. With the back legs, don't nick the back leg tendon of the goat. It's best to shear goats for fiber when the weather is warmer, the goat needs to have a light layer of its mohair or cashmere to protect it from the weather. If it is cold, then you should put a goat coat on it, to protect it and keep it warm. After shearing check that there are no nicks/cuts on the goat, and if there are, use Blu-Kote or an antiseptic ointment because you don't want your goat to get an infection.

When you sell the mohair or cashmere, it needs to be as clean as possible without any debris in it, otherwise it will not be purchased. So, when you finish shearing, you'll need to pick through the fiber to remove any dirt and debris, manure tags, as well as short sections and second cuts. You can get a skirting table that will help short fibers and debris drop from the fleece. A skirting table is essentially a rectangle of welded wire attached to a wooden frame. But, even if you have one, you'll still need to pick through the fleece to remove any short sections, felted sections, second cuts, as well as manure tags and vegetable matter. When you've cleaned the fleece, you can weigh it and keep the records of how much of usable fiber you get from each goat.

Assuming you are new to fiber goats, it would be advisable to start out small, and initially one buck and three does would be a sufficient sized herd for a person new to this. It can take time to wash the goats and prepare them for shearing. Fiber goats can be more expensive than meat or milk goats. In the US, if you are classed as a homestead rather than a hobby farm, this changes your taxes and can be beneficial. You will need to report all income, but you can claim all expenses that your goats cost you too (animals, feed, vets, shearing, bedding, farm equipment, any phone/Internet used for business, work uniform, spinning/dye equipment, goat coats/feeders etc.).

Goats should be sheared twice a year because goats like Angora can't shed their coats. Other goats may shed partially and will rub on fencing or other items. Type C goats can have their fiber combed off when it starts to release.

Roving (this is the step before fiber is spun into yarn, it still has a fluffy appearance, the fiber here has been collected from the goat, cleaned, and combed out until they web and lay nicely together in the same direction) can be made from clean goat fiber and spun into yarn. You can also use mohair wet or in needle felting, or it can be woven.

Weed Control

Goats can be used to control weeds, they graze it and eat away at it, so they do not have the opportunity to grow or flower and disperse seeds. Goats often graze plants that sheep or cattle would not eat. Goats can manage to get to weeds in rocky areas, around trees, and corners that spraying has not managed to kill.

Any type of goat can be used for weed control, but it tends to be Boer goats and rangeland goats (these are goats raised on land which has grasses, shrubs, and herbs that can be browsed) that are more typically used.

The goats will constantly graze, and by eating weeds leaves, it will take away the weed's source of energy, and the goat will also trample and ringbark trees (remove the bark from trees, which will eventually kill it, it will result in death of the area above the ring of wood that has been eaten).

Using goats for weed control saves you time and money in terms of chemicals, labor, and machinery. They will efficiently and in a sustainable manner keep weeds down, it's kinder to the environment rather than using chemicals, and you have control over which areas your goats graze in. You can start growing other pasture in the area which will outgrow weeds, and cover any bare patches, clover is a good choice for this. It is good to use fertilizer to help the clover grow.

Before you set goats loose on weeds, it's sensible to have an idea of what percentage of your ground is covered by weeds so that you gain an impression of how infested by weeds the paddock is. You can monitor how much weeds the goats are eating and ensure that they have enough to graze on to meet their nutrition requirements. If goats are grazing weeds, the fence enclosure needs to be safe, and they also require water points too. If goats are being let into wooded areas to clear weeds, it's still best to isolate the area you want them to graze in with fencing. If the weed infestation would require too many goats to deal with, you can do integrated control, where some weeds are slashed, cultivated, sprayed, and cropped before the goats come in. If the area has a lot of blackberry or gorse, then paths may need to be cut through these to allow the goats to have good access. To fully eradicate weeds, it may need several years of goats controlling the weeds. Some people choose to opt for a three level of control, the first being spraying, the second sheep grazing who will eat the vegetative weed, then goats will eat the high fiber material and inaccessible weeds.

Blackberry is something that goats love and can graze all year round, so is Scotch broom, sweet briar, and wild turnip. Thistles tend to be most palatable to goats when they are flowering. If goats have grazed an area, you can burn any remaining canes in late summer. If you don't want goats to graze on the bark of a tree and kill it, then place chicken mesh around the trunk of the tree, up to a height of around 6-7 feet. Goats will even eat things like poison ivy. If you are growing flowers like roses, however, do ensure you keep the goats away from them.

Many places will rent/hire goats to clear brush along roadsides, lots, playfields, and parks. So, if you have a herd of goats, this is one way that you could choose to make money from them. Goats will eat kudzu, oriental bittersweet, sumac, ironweed, winged elm, mile-a-minute, ailanthus, and stinging nettles. Goats can be poisoned by eating yew, so don't ever allow them to eat this. Also poisonous to goats are horse nettle, poison hemlock, mountain laurel and Jimson weed. If you want goats to clear a large area, 30 goats will clear an acre of brush and weeds in three days. But, if your home area is much smaller, you will need fewer goats than this. You will need to supplement goats' food at home with other things like alfalfa hay. Whilst goats may nibble some grass, they are not good for keeping lawns in check, they prefer weeds and shrubs to graze on.

It costs 50% less to hire goats to clear land than it would to hire labor or rent machinery. They will climb and navigate areas that a tractor couldn't get to. Goats cannot be trained to eat some plants and not others, so if there are plants you want protecting, ensure these are fully goat-proof.

When goats are eating weeds, they are clearing unwanted plants, and turning these into milk, meat, and manure. I'll talk about manure more in the next section, but clearly, this stays on the land cleared by the goats and enriches it for whatever is planted next.

Manure

Goat manure is fantastic to put in garden beds to grown plants. Goat manure is dry pellets that are easy to collect and apply without the mess created by other manure. Goat manure can be used for flowers, vegetables, herbs, and trees, it can be put in compost to be used as mulch too.

The manure doesn't attract insects or burn plants (which can be issues with cow and horse manure). Goat droppings need to be composted before use.

If you're gathering the compost for use yourself on your own land, an ideal time to do this is when cleaning out your goat shelter. When you gather the goat droppings and some wet urine-soaked straw or wood shavings, ensure that you have at least 50% manure, and place this in a compost device or pile somewhere in your garden. You can add leaves and shredded newspaper too, in order to add some carbon.

To compost goat manure, simply place it into a composting device and add things like grass clippings, leaves, kitchen scraps, eggshells, tea bags, and so on. It is good to keep the compost moist and stir it if possible.

By composting goat manure, microorganisms digest organic matter and covert these into nutrients, which makes valuable compost. Moisture is important for good goat compost; it should be 50-60% moisture. Compost should ideally be turned 2–4 times a day in the first week, then once every two days in the following weeks.

Goat manure will enrich soil and improve its condition and texture. Goat manure also doesn't have an odor, which is a huge benefit. If you are composting it (which is recommended before use) you should compost it for 4–6 months before adding it to soil. After compositing, it will contain 22 lb of nitrogen per ton (cow manure would have just 10 lb in a ton). If it is a new garden, you can put about 8 inches of goat manure on it, then each year do just a further 1–2 inches.

Goat manure has higher nutrients than farmyard manure. Goat manure is drier with a more balanced pH level. Goat manure doesn't attract maggots or flies (like chicken manure can). Goat manure can contain some weed seeds, but by composting it, it will reduce this issue.

If you get to the stage where you have large goat herds and a lot of compost, you can hire machines to help you do this, but initially you may start out doing this with hand tools. It is possible to bag goat manure for commercial use, and to create goat manure pellets by crushing the manure and making pellets with a machine. The pellets then need to be dried in a rotary drum before packing. It is possible to have the pellets coated with microbial agents to make bio-organic fertilizer. Goat manure should be spread onto garden soil about 6 months before planting, so the ideal time of year to do this is in early fall.

Goats as Pack Animals

Goats have been used as pack animals since the 1970s to carry supplies for scientists, and later equipment for tourists and hikers. Nowadays, it can be useful with families to carry gear, or older individuals who want to go into the wilderness. Types of goats used as pack animals include Alpine goats, Saanen, Oberhasli, Toggenburg and Lamancha. Sometimes Boer goats are used too. Goats can browse plants en route which means they don't need much food packing, and they don't require as much water as some other animals. An adult pack goat can carry $\frac{1}{4}$ of its weight and walk up to 12 miles a day. So, if a goat is 180 lb, it will carry approximately 45 lb. There are specific goat saddle bags to place on the goat to hold your belongings. Pack goats do need to have been socialized to be around humans, and this starts when goats are kids. A pack goat will not mind having the saddles placed on them. Goats naturally herd and will develop pack leader naturally too. Goats will follow the boss goat, usually in a set order each day.

Pack goats are used by hunters on big game hunts. If you're buying pack goats, you want them healthy and used to trail life and packing. Goats are cheap to purchase as pack animals, compared to horses. It is no use buying pygmy goats as a pack animal, but dairy goats are good because they have long legs and big frames. It's better to buy a wethered goat, because you don't want does carrying milk over rough ground, or an uncastrated buck that may be aggressive and smelly. A goat will eat about ¼ of what a horse would each, you can feed goats compressed alfalfa hay cubes. A pack animal should be trained on trails and hiking prior to taking them on long trips, initially you could keep the panniers empty and gradually increase the weight, ensuring it is balanced each side. Hoof care is important for all goats, and pack goats especially. They need to be able to walk comfortably, and untrimmed hooves can make them lame on their feet. A smart idea is to place boulders into goat pens because they like climbing things, this will keep their hooves filed. Pack goats can make it easier for you to go into the backcountry and stay there longer, by having the equipment you need at hand, carried by them. They can also carry back any meat that hunters harvest in game bags.

When you have reached a place of camp with goats, they can be tethered on a low line or allowed to browse free-range. You could transport the goats to the trailhead using a pickup truck. Pack goats don't tend to frighten game, and they're much cheaper than horses, mules, or llamas.

Goats as Pets

As already discussed in this book, goats can be used for meat, milk, fiber, as pack animals, for weed control, and manure, but it is important to realize that goats make fantastic companion animals too. Goats have incredible personalities, they are individuals, some are stubborn, some are mischievous, some are adventurers, they are quirky and hilarious. Pygmy and Kinder breeds make perfect companion goats. Many people, including those who live in city environments, have pet goats. Any breed of goat can make a good pet, but the smaller breeds are more suited to places that may not have a lot of land, for example, if you only have a back garden.

Pygmy goats are very popular and incredibly cute. Pygmy goats are between 16 and 23 inches high and weigh between 40 lb and 80 lb. Pygmy goats breed all year round, so you can get a pygmy kid at any time of the year. Kinder goats are good for companion goats too, a Kinder is a cross between a Pygmy and a Nubian goat. Kinder goats can be bottle-fed when they are small. They are friendly and because of their compact size they can be transported in a dog crate. They like the company of humans and enjoy walking, hiking, and camping. They are full of fun and are very entertaining to watch.

Goats get on well with other animals and can be companions to dogs. Some goats may be therapy goats at schools, assisted living places, and community centers.

Goats are very popular in petting zoos because of their friendly nature. Goats do need at least a partner, because they're herd animals, it isn't fair to just get a single goat. They also need a yard to roam around in. Some goats live for 15–18 years, so you do need to commit to keeping them. Goats like to be petted by their owners and will enjoy eating food from your hand. Goats can become jealous and aggressive if one goat is favored more than the other. Ideally, it's best to live on a rural farm or have a home with acres of land to have a goat. If you do live in a city, the laws may not allow you to keep a goat, as they are classed as an agricultural species, so do check this thoroughly before you purchase a goat. If the laws allow you to have goats and you have neighbors living close by, it's worth checking with them that they are OK with you having goats, as goats can be quite loud with their bleating.

We use our goats to clear land on our homestead and keep the weeds under control. We much prefer letting the goats deal with this naturally, rather than using any chemical sprays. The goats do a great job of clearing weeds, and it allows them to browse while giving them some variety and fiber to their diet at the same time. They've been wonderful, and as we had goats anyhow, it made sense for them to do this, rather than pay to hire labor to do this job, it gives the goats exercise, activity, and a sense of freedom too.

Key takeaways from this chapter:

1. You don't get that much wool per goat per year. A cashmere wool goat will generally give you 6 ounces of cashmere wool per year.
2. First clip, (first shearing) mohair is the softest, and should be kept separate from older goats' wool.
3. It's best to have white goats' wool, because then it can be dyed easily. Unless you're specifically selling to natural wool buyers.
4. Guard hair (the coarse hair) needs to be removed by hand from the wool you've gathered.
5. Wash the goat and thoroughly dry them before shearing. Apply antiseptic to any nicks after shearing the goat.
6. Protect any plants with chicken wire that you don't want goats to browse on and eat.
7. Be aware of what plants are poisonous to goats so that you don't allow them to browse on these and become ill.
8. Goat manure needs composting, and it's excellent to use on any part of your garden.
9. Goats can make useful pack animals, provided they've been socialized with humans, and have been given training prior to setting out in the wilderness.
10. Goats make lovely pets, especially pygmy goats.
11. It's always best to have at least 2 pet goats to keep one another company.
12. If you live in a city, check the local laws to make sure you are able to have goats. You can also check with your neighbors to see if they don't mind you having a goat.

The next and final chapter moves on to look at what products can be made from goat's milk, and some recipes with goat meat.

Chapter 12: Goat Milk and Meat Recipes

So, you've diligently followed all the steps in the book so far regarding getting goats, housing them, providing adequate fencing, feeding them a nutritious diet, taking good care of their health and well-being, breeding them, milking them, and harvesting goat meat. You may now wonder what you can do with the produce? This chapter is here to help with details about cheese making, other milk products, and some goat meat recipes to get you started.

Cheesemaking

We tend to drink the freshest goat's milk, and we have a supply that just continuously renews itself. We use day-old milk for cooking and cheesemaking. If you have milked your goat, and you live in the US, you need to decide whether you will sell the cheese you make from it, or just have it for your own use. If you decide to sell it, then you need to either pasteurize your milk or let the goat cheese age for 60 days before you can sell it.

Goat cheese, also called Chevre, is super easy to make at home, and on top of that, it's ridiculously cost effective. With just 3 ingredients, you can make a batch of lovely Chevre for your family and friends to enjoy, or to sell it.

Goat cheese is made via acid/heat coagulation, which is the oldest method of cheese making in the world. Essentially, lemon juice or white vinegar break apart the protein structure of the milk once it's reached a certain temperature.

Here are the products you'll need to make a pound of Chevre:

- 1 quart of fresh goat's milk
- 1/3 cup of lemon juice or white vinegar
- ¼ teaspoon non-iodized salt

And the equipment:

- A pot to boil milk
- A colander to drain the curds
- Digital thermometer
- 2 cheese cloths

Once you've gathered the ingredients and the equipment, it's time to start making cheese. Here are the steps you have to follow to make your own lovely goat cheese:

1. Slowly heat the milk to 180°F–185°F, use the thermometer to check the temperature of the milk. The surface should look foamy and the bubbles will be forming. Turn off the heat once the milk reaches the desired temperature.
2. Stir in the lemon juice or white vinegar with a long-handled utensil. Let this mixture sit for 10 minutes. The milk will start to curdle and should become thicker on the surface.
3. Line a colander with 2 damp cheese cloths and suspend it over a large pot or bowl. Pour the milk gently into the colander lined with cheese cloths. Cover and let it drain for 2 hours.

4. Gather the cheese cloth up around the curds and tie it into a bundle. You can use a rubber band to tie it up. Hang the bundle over a pot so that the liquid can drain. You can do it by tying the bundle to a long utensil and putting it over a pot. Let it drain for 4–8 hours. The longer you let it drain, the thicker the cheese will be.
5. Once the liquid is drained, untie the bundle and put the cheese into a bowl. Mix in the salt to taste.
6. Form the cheese into a log or a wheel with your hands, just like you would form batter. You can use a cookie cutter to mold cheese as well. You can use the cheese right away or refrigerate it for up to 2 weeks. Enjoy!

Dairy products

Yogurt

You can easily make goat milk yogurt following a simple recipe. Because goat's milk has low lactose levels, it's sometimes a better option for those with lactose intolerance. You can purchase a yogurt maker or use a cooler that is big enough to contain two 1-quart jars. You will also need a candy thermometer. Here's what you'll need:

- 2 quarts of pasteurized goat's milk
- 2 teaspoons of gelatin
- ½ cup of plain yogurt with live culture or 1 five-gram packet of freeze-dried yogurt starter
- Yoghurt maker or 9-quart flip-lid cooler
- Candy thermometer

Method:

1. Pour ¼ of the milk into a bowl and add the gelatin, don't stir it, just let it settle on the surface, set this aside.
2. Put the rest of the milk into a pan and put in the candy thermometer. Heat the milk to 180°F, then reduce the heat to low and stir for 20 minutes.

3. When the milk reaches 120°F add the gelatin mixture. When the milk reaches 108°F place it in an ice bath. Add the yogurt starter culture or powdered yogurt culture. Mix thoroughly but gently.

4. Fill the cooler halfway with boiling water. Put the cultured milk into two 1-quart jars, seal the lids and place them in the water. Close the cooler or yogurt maker and allow it to incubate for 5–10 hours. The longer it is left to incubate, the firmer the yogurt becomes, and the tarter.

5. Refrigerate the yogurt until you use it, only keep it refrigerated for up to a fortnight, then discard.

Kefir

Kefir is a lovely refreshing fermented milk drink. Everyone in my family loves kefir. I used to drink it since I was a little kid, and I absolutely love it to this day.

To make kefir, you need milk kefir grains, or a kefir starter culture. Using a direct-set powdered kefir starter culture is a good option to make kefir without milk grains. All you need is a start culture and a quart of pasteurized goat's milk.

Pour the milk into a glass or plastic container, add the kefir starter culture until it has dissolved. Cover the container and place in a warm spot 72°F–74°F for 12–18 hours. Once it has finished, put a lid on it and keep it in the refrigerator. I prefer to drink it plain, but you can flavor your kefir with fruit, honey, spices, or sugar if you wish.

Kumis

Kumis is a healthy refreshing drink made from goat's milk. It has a peculiar sweet taste. It can aid digestion and restore microflora to the upper respiratory tract, which can be a great aid to health.

To make it, you need a quart of the freshest goat's milk, a glass of boiled water, 3 tablespoons of kefir, 3 tablespoons honey or sugar, and 5 grams of baker's yeast. You pour the boiled water and sugar into the milk, then put in a teaspoon of kefir into the mix at a time, it should mix without lumps or flakes. Wrap the container in a cloth, and place in a warm place overnight for 5–6 hours, then the thick substance will need filtering through a strainer. Add water to the yeast and mix this into the kumis and leave for 5–10 minutes. It can then be poured into bottles and refrigerated. After a day of it being in the fridge, you can drink it. It should definitely be drunk within 3–5 days.

Butter from Goat's Milk

Goat milk butter has a lower melting point than butter made from cow's milk. It contains more unsaturated fatty acids. Goat milk butter is pure white in color. It takes longer for cream to form in goat's milk unless you have a separator (but it's hard to find these, except second-hand).

In order to make goat milk butter, you will need the following equipment:

- Dairy thermometer
- Glass churn or a mixer
- Pan that can fit into a larger pot of water

And these products:

- Goat milk cream
- Cold water
- Salt to taste

And here's how you make goat milk butter:

First, you need to separate the cream in goat's milk. There are two ways to do it. You can purchase a cream separator, which will make the job quicker and easier. Or you can simply put a jar of milk into the fridge and refrigerate it at least overnight, or even better, for a few days. The heavy cream will rise to the top as the milk cools down. If you let your milk sit for 3–7 days, you'll get about ½ inch of cream that's risen to the top. Scoop the cream from the top with a spoon until you get to a more watery part. Try to get as little of the watery milk as possible. Once you collect enough cream, it's time to move on to making butter. I suggest getting a quart of cream. You can store it in the freezer to preserve its freshness until you gather all the cream. Don't worry, the cream will stay fresh in the freezer even if it takes you a month to gather the cream.

Now that you have the cream, it's time to move on to making butter. Here are the steps you need to follow to make your own lovely goat milk butter:

1. Once you collect the cream, let it stand overnight so that butterfat globules become ripe.
2. Heat the cream in a double boiler to 146°F, put the pan in cold water, and let it cool down to somewhere between 52°F and 60°F.
3. Pour the liquid into the churn or mixer (only half full) and start agitation. If you're using a churn, the butter should come 30 to 40 minutes after this. If you're using a mixer, mix on high for about 10–15 minutes. If the blender starts bogging down, you can add a little goat's milk to make the mass less dense.
4. Once you can see larger balls of butter forming, stop churning or turn off the mixer and scoop out the balls of butter.
5. Place the butter into a bowl of cold water and wash out the excess buttermilk by squeezing the butter with a spoon. Repeat the washing process a few times until the water is clear.
6. Once the butter is clean, squeeze any excess water from it with a spoon
7. Store the butter in the fridge. Some excess water may still come out of it during the first day—just pour it off.

And that's it, enjoy your own homemade goat milk butter! If you need it for later, you can freeze the butter for 6 months.

Homemade Buttermilk

Buttermilk is what is typically left over after making butter. Raw milk is churned, and the liquid that drains off is buttermilk. Buttermilk is light because the fat remains with the butter. It can contain probiotics like yogurt. You can, however, make homemade buttermilk in 10 minutes. If you are using buttermilk in recipes, there is little difference. Homemade buttermilk recipes are acidic. If you need buttermilk, then simply add a tablespoon of vinegar to 8oz of milk, place to one side for a few minutes, then use in your recipe as suggested.

Goat Meat Recipes

Mexican Goat Baracoa Tacos

Yield: 4 Servings | Preparation Time: 15 minutes | Cooking Time: 8 hours

Ingredients:

- 3 lb of goat meat
- 4 cloves of garlic
- Dash of chipotle sauce
- 2 tbsp of lime juice
- 2 cups of broth
- ¼ cup of apple cider vinegar
- 2 bay leaves
- 1 tsp of cumin
- 1 tsp of ground cloves
- 1 tsp of oregano
- Salt and pepper to taste

To serve:

- Corn tortillas
- Cilantro chopped
- Onion diced
- Jalapenos slices
- Cotija cheese

Instructions:

Combine all the ingredients into a slow cooker or crockpot. Ensure that the meat has been cut into small cubes and is covered in the liquid. Cook on low for 8 hours. Take the meat out and shred it with two forks. Discard the bay leaves. Spoon the braising liquid back over the shredded goat meat, then place in tortillas, and top with the cilantro, onion, jalapenos, and cotija.

Jamaican Curried Goat

Yield: 8 Servings | Prep time: 30 Mins | Cook Time: 2 Hours

Ingredients:

- 1/4 cup vegetable oil
- 6–8 Tbsp curry powder
- 1 Tbsp allspice (see step 1)
- 3 pounds goat meat
- Salt
- 2 onions, chopped

- 1–2 habanero or Scotch bonnet peppers, seeded and chopped
- A 2-inch piece of ginger, peeled and minced
- 1 head of garlic, peeled and chopped
- 1–2 15-ounce cans coconut milk
- 1 15-ounce can of tomato sauce or crushed tomatoes
- 1 Tbsp dried thyme
- 3–4 cups water
- 5 Yukon gold potatoes, peeled and cut into 1-inch chunk

Instructions:

1. Make the curry powder

If you can find Jamaican curry powder, definitely use it. If not, use regular curry powder and add the allspice to it. You will need at least 6 tablespoons of spices for this stew, and you can kick it up to 8–9 depending on how spicy you like it.

2. Cut and salt the goat meat

Cut the meat into large chunks, maybe 2–3 inches across. If you have bones, you can use them, too. Salt everything well and set aside to come to room temperature for about 30 minutes.

3. Heat the curry powder in oil

Heat the oil in a large pot over medium-high heat. Mix in 2 tablespoons of the curry powder and heat until fragrant.

4. Brown meat in curried oil

Pat the meat dry and brown well in the curried oil. Do this in batches and don't overcrowd the pot. It will take a while to do this, maybe 30 minutes or so. Set the browned meat aside in a bowl. (When all the meat is browned, if you have bones, add them and brown them, too.)

5. Cook onions, habanero, ginger, and garlic

Add the onions and habanero to the pot and sauté, stirring from time to time, until the onions just start to brown, about 5 minutes. Sprinkle some salt over them as they cook. Add the ginger and garlic, mix well and sauté for another 1–2 minutes.

Put the meat (and bones, if using) back into the pot, along with any juices left in the bowl. Mix well.

6. **Add coconut milk, tomatoes, curry powder, water, thyme, then simmer**

Pour in the coconut milk and tomatoes and 5 tablespoons of the curry powder. Stir to combine. If you are using 2 cans of coconut milk, add 3 cups of water. If you're only using 1 can, add 4 cups of water. Add the thyme.

Bring to a simmer and let it cook until the meat is falling apart tender, which will take at least 2 hours. Longer if you have a mature goat.

7. **Add potatoes**

Once the meat is close to being done—tender but not falling apart yet—add the potatoes and mix in. The stew is done when the potatoes are. Taste for salt and add some if it needs it.

8. **Skim fat**

You might need to skim off the layer of fat at the top of the curry before serving. Do this with a large, shallow spoon, skimming into a bowl. Also, be sure to remove any bones before you serve the curry.

The stew is better the day after, or even several days after, the day you make it.

Serve with Jamaican rice and peas, a coconut rice with kidney beans.

Goat Shawarma

Yield: 4 Servings | Prep Time: 30 minutes | Cook Time: 2.5 hours

Ingredients:

- 2 lb of goat meat
- 4 garlic cloves minced
- 1 teaspoon cumin
- 1 teaspoon black pepper
- 1 teaspoon coriander
- 1 cardamom pod
- ½ teaspoon paprika
- ½ teaspoon nutmeg
- ½ teaspoon cinnamon
- ½ teaspoon cloves
- ½ teaspoon turmeric
- 1 cup of lemon juice
- ¼ cup vinegar
- ¼ cup of olive oil
- Salt to taste

To make tahini sauce:
- ¾ cup tahini paste
- 1–2 garlic cloves
- ½ tsp salt
- ½ cup lemon juice
- ¼ cup cold water

To serve:
- Pita breads
- Pickles
- Shredded lettuce
- Tomatoes

Instructions:

1. Make the shawarma spice mix by toasting the spices gently until fragrant. Do this the day before, then grind the spices in a pestle and mortar. Add the shawarma seasoning to salt, vinegar, lemon juice, olive oil and minced garlic. Marinate the goat meat overnight.

2. The next day, preheat the oven to 275°F, place the meat on a tray and cook for 2 hours. Once the meat is tender, remove and let it cool for 10 minutes.

3. Make the tahini sauce by mixing tahini with lemon juice, garlic, salt and water

4. Griddle the steaks until brown and crispy, then cut into slices after resting 10 minutes.

5. Place the goat shawarma in pitas with shredded lettuce, tomatoes, pickles and the tahini sauce

Wholemeal Penne with Kale and Goat Cheese

Yield: 4 Servings | Prep Time: 5 minutes | Cook Time: 20 minutes

Ingredients:

- 500 grams of wholemeal penne
- 1 tablespoon of olive oil
- 1 bunch of curly kale
- 2 onions
- 150 grams of goat's cheese
- ¼ bunch of basil leaves
- Salt and pepper to taste

Instructions:

1. Boil pasta in a pan of salted water until al dente.
2. Whilst pasta is boiling, chop onion and fry in a pan until soft and golden.
3. Chop the kale and stir into the pasta for 5 minutes.
4. Season with salt and pepper.
5. Add in goat cheese and basil.

We have long been experimenting with goat meat and different dairy products, such as cheese, yogurt, butter, and other products. Goat produce is so versatile and has such a delicious, distinctive taste. We've converted numerous family and friends by inviting them round and letting them try some of our meals at dinner parties. Some of my favorites are slow-cooker dishes because they're so quick and easy. Literally 10 minutes prep throwing all the ingredients into the pot, then the next day after a long day working on the farm, you come in to a truly delicious meal, where the goat meat just melts in your mouth.

Key takeaways from this chapter:

1. If you decide to sell the goat cheese that you make, then you need to either pasteurize your milk or let the goat cheese age for 60 days before selling it.
2. You can make yoghurt, kefir, and kumis from goat's milk, with some starter packs.
3. You can add vinegar to goat's milk to create buttermilk for recipes.
4. You can make goat milk butter a few different ways, it will be very white, unless you put coloring into it.
5. There are numerous delicious recipes you can make with goat products. I only included my 5 favorite recipes here, as this is a guide to raising goats, and not a cookbook. But if you want to discover more recipes with goat meat, milk, and cheese, you can download the free bonus cookbook with 20 delicious recipes. You can find the link by going back to the beginning of the book right before the table of contents.

Conclusion

Buying goats is one of the best decisions I have ever made in my life. They have enabled me to spend time with creatures that I love in a very outdoorsy lifestyle that I love. We have created a sustainable and profitable livelihood for ourselves based around goats. The majority of the meat, milk, cheese, yogurt, butter and ice cream that we consume is produced by our own goats and the work we put in to make this happen on our smallholding. We also produce make our own jumpers, hats, and scarves from the fiber we shear from our goats. We have goats tackle weeds on our property and use composted goat manure to enrich the soil and make our plants and flowers healthy. It's a life that I love and I would encourage those who love goats to pursue their dream.

So, this book has taken you through a complete guide to raising goats for beginners looking at breeds, housing, feeding, goat care, breeding, dairy, and meat. I have tried to include all the information that I've picked up over the years, and often learned the hard way, that I wish someone could have imparted to me at the start of my goat-owning journey. I want to save you time and to save you the mistakes I made. I want you to feel supported in knowing and understanding how to best care for your goats. Any tips and tricks that I've learned throughout decades of keeping goats, as well as chatting to other goat owners, I've included in here to help you.

Chapter 1 covered the benefits of keeping goats and gave an in-depth explanation of the most well-known breeds for milk, meat, and fiber goats; there is a glossary of terms to do with goats, that when you're first starting out may be baffling, until you start to speak the language of goat owners too. I also dispelled some common myths about goats.

Chapter 2 explained where you can purchase goats from and gave you an idea of the cost of goats, including the hidden expenses that you may not consider initially. The chapter gave advice on how to get a good, healthy goat and how to transport them back to your home.

Chapter 3 looked at the shelter that goats require, the size of the shelter according to how many goats you have, the bedding they need, and the food troughs they require.

Chapter 4 looked at the subject of fencing which is super important to keep your goats safe, but be warned the fencing may have to be higher and stronger than you first anticipated, as goats will test the fencing, and be prepared to do quick repairs as and when needed.

Chapter 5 looked at what you should feed your goats and what foods to avoid. Goats should always have access to fresh water. If you feed your goats with hay, they should have supplements and minerals too.

Chapter 6 looked at the other things that you need to do to care for your goats, from getting their hooves trimmed, disbudding and dehorning, tattooing, and shearing goats.

Chapter 7 covered goat health and how to find and use a veterinarian. In this chapter, there is a list of common health issues that you need to check your goats for and contact a vet if they show any signs of them.

Chapter 8 is about breeding goats, how to prepare for breeding, caring for pregnant goats, successful birth, newborn checks, and raising kids. Having newborn kids on the farm is one of the nicest parts of goat ownership, in my opinion.

Chapter 9 focuses specifically on milking goats, the equipment you require, the process of milking goats, and then how to handle the milk so that it's safe for human consumption and can be sold to customers.

Chapter 10 looks at harvesting goat meat, covering when and how to butcher a goat, though I would recommend you take your goats to a butcher. The chapter looks at meat quality and how to store goat meat.

Chapter 11 looks at some additional benefits of goats. Another favorite part of goat ownership for me is having goats for fiber. Other benefits include goats clearing weeds, and using goats as pack animals to carry items for camping and hiking trips, and of course goats as pets or companion animals too.

Chapter 12 looks at how easy it is to make goat cheese, butter, yogurt, kefir, kumis, and some delicious goat meat recipes.

My advice would be to read through this book in chronological order because it takes you through a logical journey of learning more about goats, and what they need to thrive, to specific types of goat for milk, meat, and fiber. It's a book which you can keep and refer back to, once you own your own goats too, to give advice on feeding, care, health, and recipes. I would really carefully consider the budget you have available to buy everything you need to care for goats, then if that all looks possible, start preparing your land and pastures ready with fencing and housing in preparation to bring your goats home. I hope that they bring you as many years of pleasure as ours have brought us.

Herd Management Calendar

While this book contains all the necessary information to keep your herd healthy, it's nice to have all things consolidated in one convenient place. You can find various goat herd management calendars online. I personally find the one by American Dairy Goat Association to be one of the most comprehensive and versatile.

You can use this goat herd management calendar as a guide to assist you in preparing for each season. Some breeds and breeders may have unique needs or practice out-of-season breeding. Keep in mind, this is just a sample calendar, it can and should be adapted to your needs. You can this calendar as a starting point and then tailor it to your needs gradually. Alternatively, you can find a lot of other herd management calendars online.

I always suggest seeking the advice of your small ruminant veterinarian and never disregarding professional advice or delay seeking professional veterinarian assistance. With that said, please find the calendar below:

SPRING

Prepare for kidding

- Have kidding area cleaned and bedded with fresh straw several days before the doe's due date.

- **Get supplies ready:**
  - A good light in the delivery area.
  - A clean bucket for water.
  - Surgical scrub, such as Nolvosan, or a bottle of mild detergent (e.g. Dawn, Ivory, Joy) for cleaning hands and the vulva of the doe.
  - Obstetrical lubricant (Lubrisept, K-Y) and, if possible, disposable obstetrical gloves for assisted births.
  - Dry towels for cleaning and rubbing kids.

- Iodine (7% tincture) for dipping navels. A small jar or film canister for individual use is handy. Dip navel immediately after birth, and repeat in 12 hours.
- Scissors and dental floss for umbilical cord.
- Keep frozen colostrum from a safe, CAEV-free source. To heat-treat colostrum, heat colostrum to 135°F in a double boiler or water bath and maintain temperature for one hour.
- Clean bottle and nipple for feed-ing colostrum.
- Feeding tube (12–18 French) and large syringe (35–60 cc, with cath-eter tip) for giving colostrum to weak kids.

Disease Prevention

- Tape doe's teats one week before due date with teat tape. This will prevent kids from possibly nurs-ing a transmittable diseased doe.
- Segregate disease-positive parturient does from the rest of the herd to prevent horizontal transmission from infected genital secretions.
- Remove kids from doe immediately after birth.
- It is advised to bathe each kid in warm water with a mild detergent (e.g. Dawn, Ivory, Joy) to remove any vaginal secretions from the doe. Thoroughly dry kid with a warm hair dryer until completely dry.
- Feed colostrum from a safe source within the first couple hours after birth. Give 10% of kid's body weight within 18 hours (e.g., 13 oz for an 8 lb kid). Then feed pasteurized milk, disease-free milk, or milk replacer.

Nutrition for the Doe

- Have pregnant does on a rising plane of nutrition in late gestation, i.e., good quality grass hay, supple-ment with some leafy alfalfa. Grad-ually increase grain ration in last few weeks to provide energy.

- Work with your veterinarian or livestock nutritionist about increased energy and calcium needs during gestation.

Disease Prevention: Does

- Be sure does are boostered for CDT in last 4–6 weeks prior to due date. Consult your veterinarian for advice on selenium supplementation for does and kids in deficient areas.
- Deworm doe 1–2 weeks postpartum.

Disease Prevention: Kids

- Begin Coccidiosis preventive or start monitoring fecals by three weeks of age.
- CDT series at 4, 8, and 12 weeks of age.
- Begin strategic deworming at 6–8 weeks.

SUMMER

- Be sure kids receive their CD-T boosters (e.g., 8–12–16 weeks).
- Wet weather has given parasites a big boost in many areas. Practice strategic helminth (worm) control in all groups of animals. Doses of deworm-ers in goats are usually 2X the cow or sheep dose (4X the cattle dose for Fenbendazole-PanacurR). In the case of Ivomec, use the oral formulation. Resistance to all dewormers is appearing, so monitor success with quantitive fecal exams. (See Best Management Practices to Control Internal Parasites in Small Ruminants articles)
- Rotate pastures every several weeks or allow forage to grow to 6-8" tall before reintroducing animals. Another common practice is to allow another species to graze the pasture while goats have been rotated off.
- Coccidiostats for kids.
- Check for external parasites; keep animals clipped and clean.
- Be careful with grain overload dur-ing peak lactation, and when get-ting ready for show. Increases in concentrate feed must be made gradually, over a couple of weeks.

- Be sure fresh water is always present. Consumption goes way up in warm weather, and during lactation.

- Monitor presence of poisonous plants which may have grown within reach of animals.

- When hauling in hot weather, provide good ventilation. While traveling, will animals have fresh air and water?

- At show time, be careful not to "over-udder" a doe as she can develop an allergic reaction to backed-up milk under pressure and be at risk for developing mastitis.

Pre-breeding Buck Preparation

- Administer Vitamin-E/Selenium in Selenium-deficient areas.
- Keep feet trimmed.
- Offer a diet of forage and increasing amounts of concentrate in late summer.

FALL

The Buck

- Check and trim feet. Treat hoof rot as necessary.
- Check teeth on older bucks.
- Shorten or remove scurs prior to breeding season.
- Clip belly. Examine penis and prepuce for injuries and inflammation.
- Check general body condition. Improve nutritional status if too thin.
- Perform fecal and deworm as needed.
- Bo-Se in selenium-deficient areas.

The Doe

- Check and trim feet before rainy season.
- Correct body condition before breeding, especially if she is too fat. Fat around the ovaries may cause poor fertility. In general, corrections in body condition (too thin, too fat) are easier and safer to make before the doe is dried off.
- Bo-Se in selenium-deficient areas.

- Perform milk cultures to pick up subclinical mastitis. (Contact your testing lab for specific instructions.)
- Consider dry-treating the herd, where mastitis has been a persistent problem.

The Herd

- Offer good quality loose minerals.
- Check fecals in different age categories (does, kids) to evaluate parasite loads. Treat accordingly.
- Consider fall strategic deworming, coming off summer pasture.
- Disease Testing: Kids over 6 months old, new additions to the herd, any animals of questionable value or con-dition.
- Cull animals of questionable value or con-dition to reduce feed costs and maximize indoor space for the winter.

WINTER

- Pregnancy check does early enough to be able to rebreed this season if open.
- Booster vaccinations (Clostridium perfringens C & D, and Tetanus) in mid- to late-gestation at least 4 to 6 weeks prior to kidding. This pro-motes high colostral antibody levels at parturition.
- Booster Vitamin E-Selenium in mid- to late gestation, in Selenium deficient areas. This bolsters uterine muscle tone and helps prevent uterine inertia and retained placentas.
- Get does into their desired body condition while they are still milking; e.g., if too fat, gradually reduce grain before drying up. There will be fewer problems with pregnancy toxemia if weight changes are made while doe is still metabolically active.
- Pregnant does should get plenty of exercise. Fit and trim does are easier to freshen, less susceptible to pregnancy toxemia.
- Keep an eye on geriatric animals for weight loss and chilling.
- Routine hoof care for all animals.

- Monitor for external parasites (lice) during this period where animals may spend more time indoors with less sunlight.
- Eliminate moldy feed.
- Get to know and enjoy your animals better during this slow time!

Resources

Organizations

American Dairy Goat Association (ADGA)

https://adga.org

American Goat Federation (AGF)

https://americangoatfederation.org

American Goat Society (AGS)

https://americangoatsociety.com

Alpines International Club

https://alpinesinternationalclub.com

American LaMancha Breeders Association

https://lamanchas.org

International Nubians Breeders Association

https://www.i-n-b-a.org

International Sable Breeders Association

https://sabledairygoats.com

National Saanen Breeders Association

https://nationalsaanenbreeders.org

National Toggenburg Club

https://nationaltoggclub.org

Nigerian Dairy Goat Association

http://www.ndga.org

American Nigerian Dwarf Dairy Association

https://www.andda.org

Oberhasli Breeders of America

https://oberhasli.webs.com

Find a Goat Vet Near You

AG Service Finder. Find a Goat Vet Near You.

https://www.agservicefinder.com/goat-vet-near-me/

Online FAMACHA Certification

The University of Rhode Island. Online FAMACHA Certification.

https://web.uri.edu/sheepngoat/famacha/

Breed Information

American Goat Federation. Dairy Goats.

https://americangoatfederation.org/breeds-of-goats-2/dairy-goats/

American Goat Federation. Meat Goats.

https://americangoatfederation.org/breeds-of-goats-2/meat-goats/

American Goat Federation. Fiber Goats

https://americangoatfederation.org/breeds-of-goats-2/fiber-goats/

American Goat Federation. Pack Goats.

https://americangoatfederation.org/breeds-of-goats-2/pack-goats/

Goat Health

Bowen, Joan S. 2014. Common Diseases of Goats. *MSD Manual Veterinary Manual*. Online.

https://www.msdvetmanual.com/management-and-nutrition/health-management-interaction-goats/common-diseases-of-goats

Collins, Dr. Michael and Dr. Elizabeth Manning. 2021. FAQs. Johne's Information Center. *University of Wisconsin - Madison School of Veterinary Medicine*. Online.

https://johnes.org/goats/faqs/

Colorado State University. Enterotoxemia of sheep and goats. Online

https://extension.colostate.edu/topic-areas/agriculture/enterotoxemia-overeating-disease-of-sheep-and-goats-8-018/

Goats. 2019. Assessing the Physical Condition of the Goat. *Goats*. August 14th, 2019. Online.

https://goats.extension.org/assessing-the-physical-condition-of-the-goat/

Goats. 2019. How to Contact a Goat or Sheep Veterinarian. *Goats*. August 14th, 2019. Online.

https://goats.extension.org/how-to-contact-a-goat-or-sheep-veterinarian/

Goat World, Behavior. *Goat World*. Online

https://www.goatworld.com/articles/behavior/behavior.shtml

Hoschton Animal Hospital. 2016. Signs of Illness in Goats. *Hoschton Animal Hospital*. Jan 15th 2016. Online.

https://www.hoschtonanimalhospital.com/2016/01/15/hoschton-ga-vet-illness-goats

Manitoba Goat Association, Goats and their nutrition. Online.

https://www.gov.mb.ca/agriculture/livestock/goat/pubs/goats-and-their-nutrition.pdf

Nicoletti, Paul. 2013. Brucellosis in Goats. *MSD Manual Veterinary Manual*. Aug. 2013. Online.

https://www.msdvetmanual.com/reproductive-system/brucellosis-in-large-animals/brucellosis-in-goats

OIE. 2021. Bovine Tuberculosis. *OIE World Organisation for Animal Health*. Online.

https://www.oie.int/en/disease/bovine-tuberculosis/

Oregon States University. Internal parasites in sheep and goats. Online.

https://catalog.extension.oregonstate.edu/sites/catalog/files/project/pdf/em9055.pdf

Pezzanite, Lynn et al. Common Diseases and Health Problems in Sheep and Goats. *Animal Sciences*. Online.

https://www.extension.purdue.edu/extmedia/AS/AS-595-commonDiseases.pdf

Purdue University Extension. Common diseases in sheep and goats. Online

https://www.extension.purdue.edu/extmedia/as/as-595-commondiseases.pdf

Thoen, Charles O. 2014. Tuberculosis in Sheep and Goats. *MSD Manual Veterinary Manual*. Online.

https://www.msdvetmanual.com/generalized-conditions/tuberculosis-and-other-mycobacterial-infections/tuberculosis-in-sheep-and-goats

Breeding and Castration

Brady, Boyd. 2019. Reproductive Management - Dairy Goats and Sheep. *Extension Alabama A&M & Auburn Universities*. 12th August 2019. Online.

https://www.aces.edu/blog/topics/sheep-goats/reproductive-management-dairy-goats-and-sheep/

Goats. 2019. Goat Reproduction. 14th August 2019. *Goats*. Online.
https://goats.extension.org/goat-reproduction/

Goats. 2019. Newborn Goat Kids. 14th August 2019. *Goats*. Online.
https://goats.extension.org/newborn-goat-kids/

PennState Extension. n.d. Reproduction. *PennState Extension*. Online. https://extension.psu.edu/programs/courses/meat-goat/reproduction

PennState Extension. n.d. Castrating Male Kids. *PennState Extension*. Online.
https://extension.psu.edu/programs/courses/meat-goat/health/castrating-male-kids

Goat Shelter

PennState Extension. n.d. Goat Housing and Facilities. *PennState Extension*. Online.

https://extension.psu.edu/programs/courses/meat-goat/basic-production/general-overview/goat-housing-and-facilities

Hobby Farms. Tips for providing your goats the shelter they need. Online.
https://www.hobbyfarms.com/tips-providing-goats-shelter-they-need/

Dairy Goats

PennState Extension. n.d. Dairy Goat Production. *PennState Extension*. Online.
https://extension.psu.edu/dairy-goat-production

Weedemandreap.com. How to milk a goat. Online.
https://www.weedemandreap.com/how-to-milk-a-goat/

Meat Goats

PennState Extension. n.d. Meat Goat Production. *PennState Extension*. Online.
https://extension.psu.edu/meat-goat-production

North Carolina State Extension. Nutritional Feeding Management of Meat Goats. Online.
https://content.ces.ncsu.edu/nutritional-feeding-management-of-meat-goats

Fiber Goats

Agricultural Marketing Resource Center. Goats for Fiber. Online.
https://www.agmrc.org/commodities-products/livestock/goats/goats-for-fiber

Goats. 2019. Goat Fiber Production. 14th August 2019. *Goats*. Online.
https://goats.extension.org/goat-fiber-production/

Index

A

abortion, 124
AG Service Finder, 108
Albendazole, 101
alfalfa hay, 48
Alpine goats, 22
American Dairy Goat Association (ADGA), 32
American Goat Federation, 22
American Goat Society (AGS), 32
Angora goats, 28, 150
antibiotic eye ointment, 118
antibiotics, 124
Artificial Insemination (AI), 48, 124
auctions, 39

B

baking soda, 119
Banamine, 119
barn, 53
bedding, 55
Benzimidazole dewormers, 101
birthing kids, 127
blood stop powder, 118
Blu-Kote, 118
Boer goats, 26
bottle feeding, 129
breeders, 39
breeding, 122
breeding season, 122
browsing, 79, 87
Brucellosis, 43
brushing, 91
butchering a goat, 146
butter, 168
buttermilk, 169
buying goats, 39

C

C&D antitoxin, 120
Canadian Goat Society, 32
Caprine Arthritis Encephalitis (CAE), 43
Caseous Lymphadenitis (CL), 43
Cashmere goats, 28, 150
cashmere wool, 150
castrating, 102
castrator rings, 103
cattle panels, 73
cesarean section, 128

chain link fence, 74

cheesecloth, 135

cheesemaking, 163

chevon, 148

chevre, 164

Chlamydia, 124

cleaning goat shelter, 57

cleaning milking equipment, 141

clippers, 152

Clostridium (CDT), 100

concentrate, 86

concrete floors, 54

costs, 45, 46, 47, 48

Craigslist, 42

Crotalaria, 87

curds, 165

Cydectin, 102

D

dairy breeds, 21

dairy goats, 22

dairy products, 165

deep litter method, 56

dehorning, 104

deworming, 100

diet, 80

Di-Methox 40%, 120

dirt floors, 55

disbudding, 104

disbudding iron, 106

disbudding procedure, 106

disposable milk filters, 135

E

elastrator, 103

electric fence, 69

F

fainting goats, 27

FAMACHA score, 101

fecal egg count, 100

fecal exams, 100

feeding, 79

Fenbendazole, 101

fencing, 65

fertilizer, 159

fiber, 150

fiber breeds, 28

first aid supplies, 118

Five Point Check, 101

floor, 54

flushing, 124

foraged food, 81

G

gates, 75

gestation period, 122

goat as pack animals, 159

goat care, 91

goat first aid kit, 118

goat health, 99

goat meat recipes, 170

goats as pets, 160

grain, 48, 83

grooming, 91

H

harvesting goat meat, 143

hay, 82

herd management calendar, 178

hoof care, 92

hoof rot, 92

horns, 104

I

Imidazothiazole dewormers, 102

Ivermectin, 102

Ivomec, 102

J

Johne's disease, 43

K

kefir, 166

kidding, 128

Kiko goats, 27

kumis, 167

L

LaMancha goats, 23

latches, 76

leash training, 96

Levamisole, 102

Levisol, 102

lime powder, 55

lime wash, 55

M

Maalox, 120

Macrolide dewormers, 101

manger, 58

manure, 157

meat breeds, 25

milking, 134

milking pail, 135

milking schedule, 136
milking stand, 47, 135
minerals, 87
mohair, 151
molasses, 120
mold, 48, 82
Morantel Tartrate, 102
Moxidectin, 102
Myotonic goats, 27

N

Nigerian Dwarf goats, 23
nightshade, 87
Nigora goats, 150
non-deep litter method, 56
Nubian goats, 24

O

Oberhasli goats, 25
open, 125

P

Panacur, 101
pasteurizing milk, 137
peach leaves, 87
pedigree, 34
PenG, 120

plum leaves, 87
poisonous plants, 87
poke weed, 87
puberty, 123
Pygora goats, 28, 150

R

Rabies, 99
raising kids, 130
raw milk, 137, 139
recipes, 163
registration, 31
Rumatel, 102
ruminant, 35, 80

S

Saanen goats, 23
Sable goats, 24
Safe-Guard, 101
Savanna(h) goats, 26
shearing, 95
shearing procedure, 152
shed, 53
shelter, 52
stall freshener, 55
storing goat meat, 148
strainer, 135

straw, 55

supplements, 86

T

tattoo, 94

teat, 136, 137

Tetanus, 100

Tetracycline, 124

thermometer, 119

Toggenburg goats, 24

toxic plants, 87

Tramisol, 102

Tuberculosis (TB), 43

Tylosin, 124

U

udder, 137

V

vaccinations, 99

Valbazen, 101

veterinarian, 108

vitamins, 87

W

water, 85

weed control, 155

wood shavings, 55

wooden fence, 68

wooden floors, 54

woven goat wire fence, 71

Y

yogurt, 165

Printed in Great Britain
by Amazon